Scientists Like Me

Stories, Advice, and Inspiration from **25 Trailblazers** with **Experiments** to Try at Home

By Kamin Science Center and JaNay Brown-Wood

Illustrated by Kristen Uroda

WORKMAN PUBLISHING
New York

To the future world changers.
You, the youth, hold the key!
—J. B. W.

Workman Kids • Workman Publishing • Hachette Book Group, Inc. 1290 Avenue of the Americas, New York, NY 10104 • workman.com

Workman Kids is an imprint of Workman Publishing, a division of Hachette Book Group, Inc. The Workman name and logo are registered trademarks of Hachette Book Group, Inc.

Design by Andrew Wang • Cover illustration by Kristen Uroda

The publisher is not responsible for websites (or their content) that are not owned by the publisher.

Workman books may be purchased in bulk for business, educational, or promotional use. For information, please contact your local bookseller or the Hachette Book Group Special Markets Department at special.markets@hbgusa.com.

Library of Congress Cataloging-in-Publication Data is available.

ISBN 978-1-5235-1678-0

First Edition August 2025
Printed in Dongguan, China (APS) 03/25

Photo Credits

Acknowledgments

This activity book is the result of the hard work, creativity, and dedication of many talented individuals, for whose contributions and support we are immensely grateful.

First and foremost, we extend our heartfelt thanks to JaNay Brown-Wood, without whom the idea for this book would have remained forever a concept. Her creativity, voice, and joy for writing for children was the microphone and amplifier for sharing the inspirational stories on the following pages.

This book could also never have become what it did without the diverse team members from the Daniel G. and Carole L. Kamin Science Center (formerly Carnegie Science Center), whose expertise and enthusiasm brought these activities to life:

- Jonathan Doctorick
- Lucile Finucan
- Marcus A. Harshaw
- Amanda Iwaniec
- Stephen M. Kovac
- Carla Littleton
- Bethany McCall
- Stuart McNiell
- Brad Peroney
- Christine Simonson
- Christina Soff
- Michaela Williams

A special thank-you goes to Gretchen Gardner, whose brilliant idea to transform one of our beloved museum programs into this book set the entire project in motion. Jason Brown, the museum's director, and Workman editor John Meils took the lead in crafting the original proposal submitted to the publisher, ensuring that our concept was brought to fruition.

This was truly a team effort. The collective vision and passion for engaging young minds in science and discovery have been the cornerstone of this book, as they are the cornerstone of everything at Kamin Science Center.

We are also grateful to our editors, Karen Smith and John Meils, and to Workman Publishing, who believed in this project and provided the platform to reach countless young readers and show them that no matter their background, they can become the next great scientists to propel humanity forward!

Finally, thank you to all of our young explorers and their families who inspire us every day. We hope this book sparks your curiosity and ignites a lifelong love for learning.

Contents

Introduction

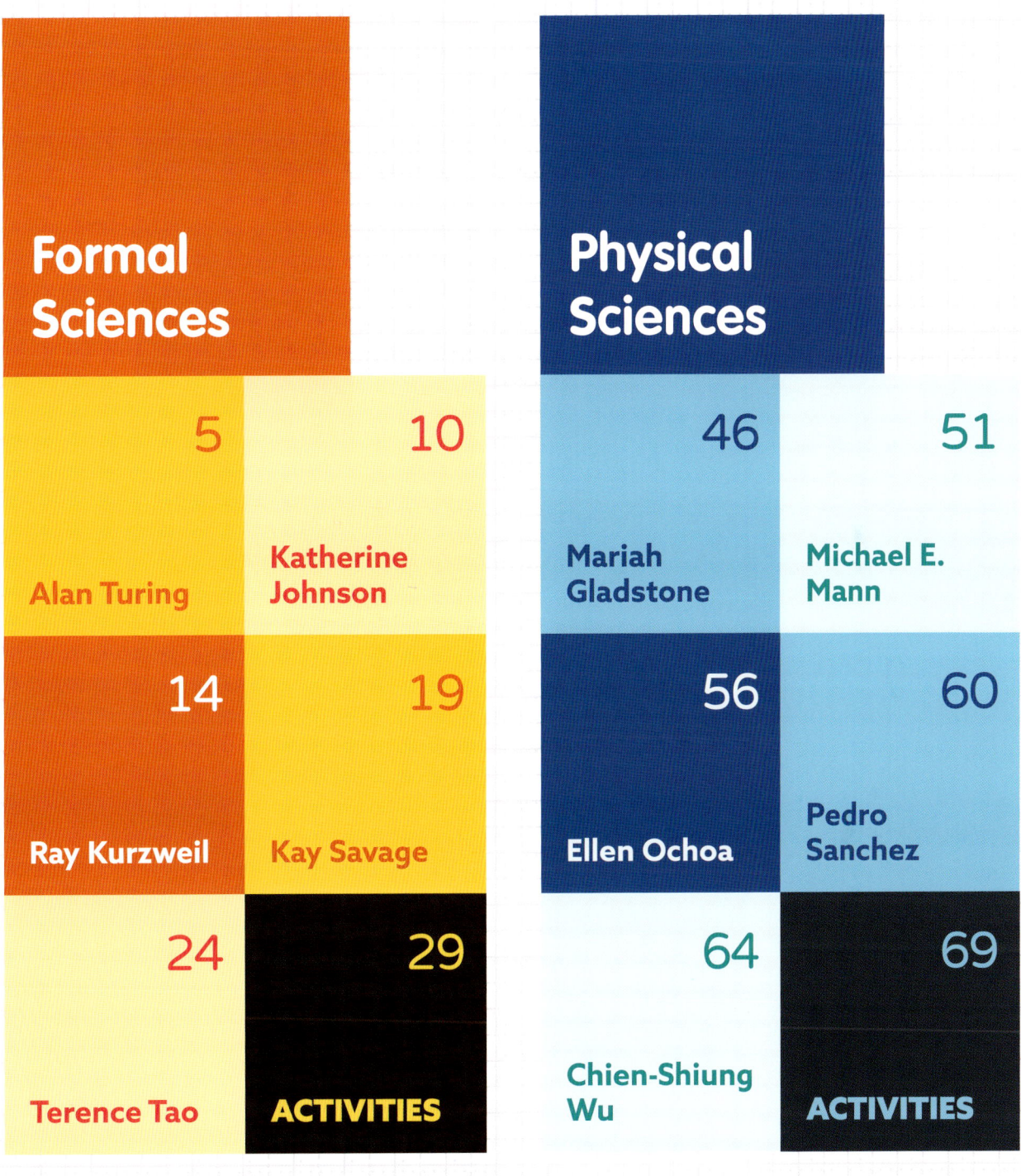

Formal Sciences

5 Alan Turing

10 Katherine Johnson

14 Ray Kurzweil

19 Kay Savage

24 Terence Tao

29 ACTIVITIES

Physical Sciences

46 Mariah Gladstone

51 Michael E. Mann

56 Ellen Ochoa

60 Pedro Sanchez

64 Chien-Shiung Wu

69 ACTIVITIES

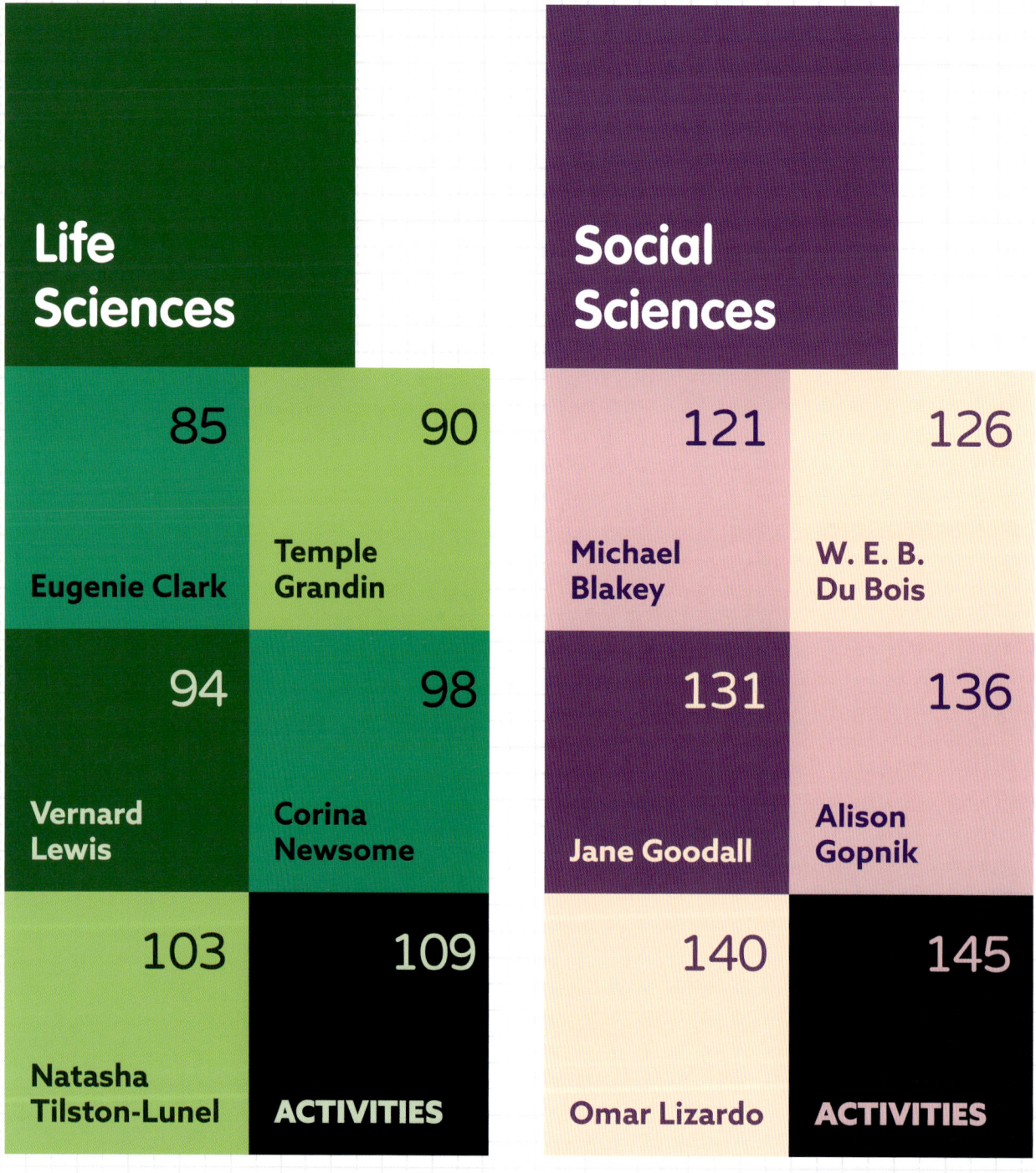

Life Sciences

85 Eugenie Clark

90 Temple Grandin

94 Vernard Lewis

98 Corina Newsome

103 Natasha Tilston-Lunel

109 ACTIVITIES

Social Sciences

121 Michael Blakey

126 W. E. B. Du Bois

131 Jane Goodall

136 Alison Gopnik

140 Omar Lizardo

145 ACTIVITIES

Medical Sciences

Explore More!

Welcome!

Hello and congratulations!

"Why the congratulations?" you may ask. Here's why: because you picked up this book! Something about the title or cover drew you in. Or maybe you had a moment to flip through the pages and skim the words or check out the pictures included within. Those pages and pictures and words and snippets of information got you thinking and wondering and questioning.

Does that sound about right?

I bet that not only did this book get you questioning, but you are probably someone who is *inquisitive* already, meaning you ask a lot of questions! Like, have you ever collected a pile of leaves in the fall and asked, "Why do leaves change color?" Or maybe you've glanced up at the sky and wondered, "Why is the sky blue?" Perhaps you've thought about people in your life and asked questions like, "Why is my little brother always bugging me?" or "Why is red my teacher's favorite color?" Better yet, you may have asked questions about animals you've seen, rocks you've found, or shooting stars you were lucky enough to glimpse jetting across the night sky.

Have you ever wondered *why* you are always asking . . . why? I know the answer! It's because you are a wonderer. You are a questioner. You are someone interested in the world around you, and you want to know where and why and which and who and how and what!

In other words, YOU have SCIENTIST written all over you! Those same questions you are wondering about are the very same types of questions that scientists inquire about and develop methods to study.

YOU are a scientist, and this book was written just for you!

So, tie up your shoes, tighten your belt, grab a notebook and a pencil, and turn your interest on high!

We've got some exploring to do!

Branches of Science

So much of the knowledge we have about everything—

spanning from the simplest elements to the most complicated and complex organisms—we have because of science and scientific thinking.

So, what is science? If we were to look at the word "science," it derives from the Latin word "scientia," which translates to mean "knowledge." And knowledge itself has to do with an awareness or familiarity, especially related to experience. This helps us understand why so much of science is built on curiosity, observation, and seeking answers about the things all around us and inside us since the start of time.

Because science spans so much, it is easier to think about science by breaking it into smaller categories, even though there is some disagreement about how scientific categories should be broken down. Should it be separated into six groups with some overlap: astronomy, biological sciences, Earth sciences, mathematics, physical sciences, and social sciences? Or should it be categorized into other types of groupings?

Although not everyone agrees about the exact way to categorize science, many do think that scientific disciplines can be separated into three overarching branches: *formal sciences*, *natural sciences (physical and life)*, and *social sciences*.

Formal Sciences

Study logic and math using systems to help us understand phenomena in the world.

Natural Sciences

The physical sciences study the inanimate world, exploring subjects ranging from rocks and glaciers to floating bodies out in the universe, while the life sciences study living things such as animals, insects, and plants.

Social Sciences

Study humans, societies, and the changes human societies experience across history, whether considering how people act in one part of the world or time in history compared to another, or why humans make the types of choices they do.

But each branch of science contains more specialized disciplines, or areas, of study. This helps scientists focus on specific areas because it would be difficult to become an expert in all areas of study. In many cases, some questions require scientists to study within more than one discipline—this would be considered "interdisciplinary," which just means more than one area of study informs the knowledge gained.

The Scientific Method

How do scientists "science"?

Even though there are so many disciplines within the three branches of science and each discipline investigates very specific types of questions, science does still include methods that are shared in just about all the science categories. This is called the *scientific method*.

The scientific method is a combination of understandings and practices that inform the process scientists use to study the world around them, helping them ask and answer questions in the most accurate way possible. Here are the elements of the scientific method:

1. **OBSERVE** *Observe* means to use your senses to take notice of the things around you. Whether it is using your eyes to look at something, your ears to listen, your nose to smell, your fingers and skin to touch, or your tongue to taste—you use your observation skills to become more aware of what is happening all around.

2. **QUESTION** After you observe, you might wonder—what, when, where, why, who, which, how? These are *questions*, and questions lead to the search for answers. When trying to understand something in the world, asking questions is often a great way to begin to think about how things work and why they are the way they are.

3. **HYPOTHESIZE** Now that you've made an observation and asked a question, you will want to make an educated guess about the answer to that question. This might mean you'll have to do more observations or read and study information related to that question. Doing some digging for valid and relevant information—also called **conducting research**—can help you make a solid hypothesis. A **hypothesis** is simply a guess that is informed by some evidence or understanding you already have. Often, you want to make sure your hypothesis is something that can actually be tested.

4. **EXPERIMENT** You've done some thinking about your observations and questions, and then made a hypothesis about them. Now's the time to experiment! This means you will decide on a series of actions to take that can help you test your question and hypothesis. Again, these steps should be informed by the knowledge you gained

from observing, reading, and studying. Your experiment will help you figure out if your hypothesis may be correct. A big part of your experiment will require you to use your observation skills again and collect information related to your experiment—also called *collecting data*. Collecting your data will give you a chance to see what your experiment revealed so you can begin to decide how close it was to your hypothesis. Collecting data may include taking notes while you observe, collecting answers from those you are studying, taking measurements of changes, and so much more! Collecting your data is important, and you want to make sure you take steps that are well-thought-out to give you a good chance of coming to a strong and accurate conclusion.

5. **ANALYZE** At this point, you must look at the data you collected and think about what it tells you. Was there a change from the start of your experiment to the end? Did your hypothesis hold true? Looking closely and thinking deeply about the data you collected helps you come to conclusions about what you found, and even about whether your experiment was as effective as it could be!

6. **CONCLUDE** Once you have analyzed your data, you should have learned something new—either something that matches your hypothesis or something that does not. Both of these findings are important!

7. **SHARE** You can use the information you learned from your experiment to share with those around you and then make needed changes to your experiment. Science is a field that always learns new things to update information and to make stronger, more accurate conclusions. Take the new information you found and share it with others, and then use it to make your experiment even better next time!

8. **REPEAT** Once your observations are done, your experiment is carried out, your hypothesis is tested, and your conclusions are made, your next steps are to repeat the whole scientific method again. You can either try the same experiment again to see if you get the same results, or make small changes to your experiment based on what you learned and try again. The more you do it, the more you learn and can share, and the more informed our world becomes!

Asking Questions

So much of scientific exploration is tied to asking good questions. Because there are so many different types of questions you can ask, it is often a great idea to try to ask specific questions. Often, when you are just starting an investigation, you might find that your questions feel more general, meaning questions that are not as narrow or focused. For example, you may ask, "What is snow?" But, as you learn more and more about your subject and do more and more research, your questions will begin to become more specific. For example,

"How do the frozen water molecules of snow crystalize and create unique snowflake designs?"

With more research, more knowledge, and more practice, you will begin to ask more specific questions. This means your question will have a growing amount of detail that helps you get to the core of what you want to study. Using detail and questioning words can help you begin your scientific endeavors!

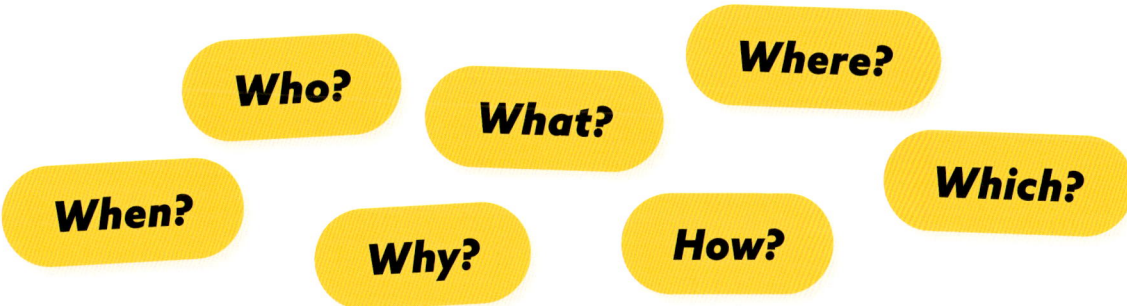

Who? What? Where? When? Why? How? Which?

What Makes a Good Scientific Investigation?

Creating specific questions that can be tested is key to making a good scientific investigation. When thinking about questions, ask yourself:

How can I test this?

What will I need to test this?

What have others done to test this question, or similar questions?

Diving In

In this book, we will explore each of the three branches of science—formal, natural, and social—and the specific disciplines within them as well. We will also explore a few interdisciplinary fields, mainly medicine, in more depth through learning about important scientists who have built their careers within these different fields of science. We will consider the various subjects they investigated using scientific methods and find out how their work has affected their fields of science and the world!

Finally, we will engage in some scientific activities within these branches. As you will see throughout this book, scientists come in all shapes and sizes, from diverse backgrounds, from all over the world—just like you—and they follow their passions to investigate all types of scientific questions and problems!

Formal Sciences

Mathematics

"What's the biggest number?"

Computer Sciences

"What makes a computer work?"

Artificial Intelligence

"Can computers think like humans?"

Information Technology

"What role does coding play in information technology?"

Statistics

"What are statistics?"

Data Science

"Can data help animals?"

Logic

"What sum will you get when you add up all the numbers from one to thirty-five?"

Formal Sciences

This area of science deals with creating and understanding knowledge based on formalized systems and, many times, without direct observation. Specific areas of science that fall within this branch include artificial intelligence, computer and data science, information technology, mathematics, statistics, and logic.

Alan Turing
1912–1954

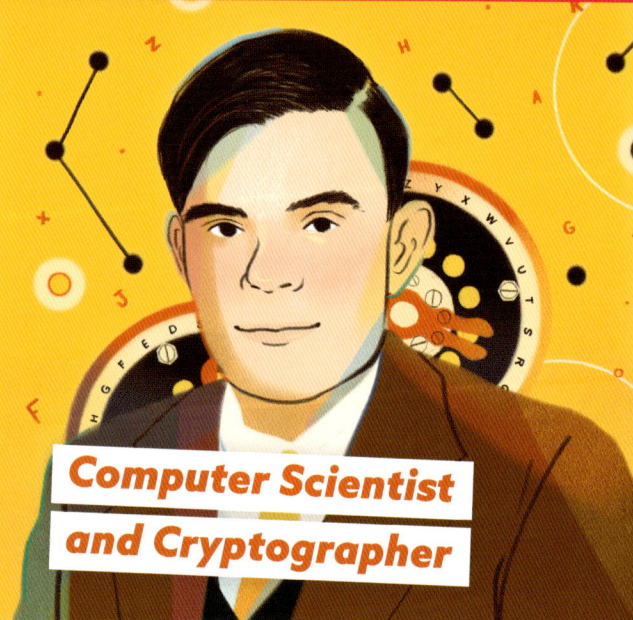

Computer Scientist and Cryptographer

Katherine Johnson
1918–2020

Mathematician and Aerospace Computer

Kay Savage

Data Scientist

Ray Kurzweil
1948–Present

Computer Scientist and Futurist

Terence Tao
1975–Present

Mathematician and Analyst

"We can only see a short distance ahead, but we can see plenty there that needs to be done."

—Alan Turing

Areas of Scientific Interest

Computer Science and Cryptography

Advice for YOU

Alan Turing liked to use math to solve applied and practical problems, like using machines to break codes. Consider your talents and skills and think about how you can use them to solve real-world problems just as Alan Turing did!

Alan Turing

Who Is Alan Turing?

Alan M. Turing was born on June 23, 1912, in London, England. His parents wanted him and his brother to grow up in England, so even though his parents had to return to work in India for long stretches of time, Turing and his brother stayed in England and lived with family friends. While Turing struggled to connect with others socially, he was really interested in the natural world—especially plants—as well as solving logical problems. As a young boy, he attended boarding schools for his education with relatively strict expectations for behavior, and Turing did not excel in his studies, especially the subjects he disliked. Luckily, at the age of ten, he received a book called *Natural Wonders Every Child Should Know*, which stoked his

passion for science. Later, he befriended a classmate named Christopher, who furthered his interest in science and math while also helping him develop better discipline and responsibility. Yes, Turing was quite talented in math and sciences, even reading a book written by Einstein and comprehending the complex theories as a teenager! Both boys planned to finish school and continue on to college together. Unfortunately, Christopher died suddenly

of an illness, leaving Turing isolated again, but he completed school and earned a scholarship to King's College at Cambridge.

During his time at King's College, Turing continued studying math, both the abstract and applied sides of it. He explored ideas related to building machines that could solve mathematical problems, and though he had not built the machines, he was able to imagine them and use the idea of them to help him learn more. Turing finished King's College, was elected a Fellow of the College, continued tinkering with the idea of machines and cryptology, and then went on to complete his PhD in mathematics at Princeton University in 1938.

During this time, conflict was brewing in Germany, and around the start of World War II, Turing joined government organizations that worked to break codes to help fight the war. He used his knowledge of machines and problem solving to develop and design code-breaking machines and, eventually, he developed a machine that allowed Great Britain to crack seemingly impossible-to-break German codes, which revealed invaluable information about the enemy such as plans and military intelligence. Because of the work of Turing, his colleagues, and his machines, Germany was defeated in the war in 1945. Some experts believe that the war would have continued for two more years, resulting in the deaths of millions more people, had it not been for Turing!

Questions for YOU

1. Turing used his mathematical skills to help end a war. What are some other ways math can be used to help those around you?

2. Do you like to break codes? What kind of codes can you break?

After the war, Turing continued his exploration of mathematics and science, even designing more machines such as the Automatic Computing Engine, publishing scientific papers, and developing new areas of science such as morphogenesis, which combined math with how natural things grow.

Despite his many accomplishments, Turing's life was filled with challenges. In addition to losing his good friend Christopher at a young age and struggling to fit in socially, Turing also faced injustices because he was gay. He was even arrested for being gay, which was illegal in Britain until 1967. After being convicted, Turing lost his clearance to do his important government work and had to endure an awful treatment as his punishment. Two years later, Turing died on June 7, 1954, of cyanide poisoning, and it was never clear whether he poisoned himself on purpose, if someone else poisoned him, or if he accidentally ingested the poison while doing his experiments. With his death, the world lost an amazing thinker and hero!

In 2009, the British prime minister issued an apology to Turing for his

Science YOU Can Explore

- **Make YOUR own code!**
  - ☐ Create some images or icons, or use numbers to stand for letters—this is your code.
  - ☐ Use your code to craft secret messages to others.
  - ☐ Give your code and the secret message to someone and see if they can break it.
  - ☐ Make it as easy or as tricky as you like!
- **Computer coding:**
  - ☐ Have you ever tried your hand at computer coding?
  - ☐ Look into your community organizations, such as the library or after-school programs, and see what types of coding programs are offered for kids your age!
  - ☐ Try it! Who knows, you may be a cryptographer in the making!

mistreatment by the British government, and in 2013, Queen Elizabeth II cleared him of his conviction. Furthermore, Alan Turing is considered to be a founding father of computer science and artificial intelligence. And even though many did not know of his heroism during the time of World War II until after his death, many of his ideas are still considered true today—like the Turing Test that strived to determine whether computers could truly think for themselves. Alan Turing's ability to explore how computers might think like humans helped establish the world of computers we know today!

?

Did YOU Know About . . . ?
Cryptology

Cryptology is the science of using secret codes to communicate information. It can include *enciphering*, or changing a message into a coded form, *deciphering*, or changing a coded message into normal language, and *cryptanalysis*, or the art of breaking codes. Codes have been used for centuries, just as the Germans did in World War II. But codes are used even today in communication, as seen in encryption used by fax machines, television, and computer security.

Areas of Scientific Interest
....

Mathematics and Aerospace Computing

Quote for YOU
....

Katherine Johnson said, "Some things will drop out of the public eye and will go away, but there will always be science, engineering, and technology. And there will always, always be mathematics."

Katherine Johnson

Who Is Katherine Johnson?

Katherine Coleman was born on August 26, 1918, in White Sulphur Springs, West Virgina—she wouldn't become Katherine Johnson until many years later, when she married her second husband, Lieutenant Colonel James Johnson. As a young child, she loved to count and she counted just about everything she came across (trees, stairs, floorboards, cracks in the sidewalk, etc.). Her interest in mathematics and her savvy as a learner allowed her to skip first grade and then fifth grade at her small, two-room schoolhouse where Black children were taught up through eighth grade. However, living in the South, she faced racism on a daily basis, and she often did not have access to the same resources as white children. This prompted her and her family to move to a different city with more opportunities.

When Katherine was in high school, she was introduced to geometry, and her passion for it was immediate. She graduated high school at the age of fourteen and began college at the historically Black West Virginia State College. Katherine took so many mathematics classes that a professor decided to mentor her and offered her even more advanced classes than the college had so she could continue to grow. Upon completing her bachelor's degree in mathematics and French, she graduated with the highest honors and began working as a teacher.

Questions for YOU

1. Have you ever thought about how math can change the world?

2. In what ways can you use your math skills to make an impact?

After graduating, Katherine heard about a center called the National Advisory Committee for Aeronautics—which later became NASA—that was hiring Black mathematicians, and she was hired for a position called a "computer." She spent her days making calculations, solving equations, and collecting her answers on data sheets that helped with the design of airplanes.

Soon, her prowess for numbers became apparent, and she was asked to join a team working to send an American astronaut into outer space—for the first time in history! With Katherine on the team, they safely blasted Alan Shepard into space, and then John Glenn after that.

Katherine continued to do amazing work utilizing her math skills and love for geometry to help others accomplish amazing feats, such as Neil Armstrong's mission to reach the Moon, and the flights of *Apollo 11* and *Apollo 13*. However, even Katherine had to overcome many challenges. Being born a Black girl in the South during the early and mid-1900s, when Black Americans were not afforded the same rights as white people, made it difficult to excel. She also faced racism and sexism in the workplace as a professional

Science YOU Can Explore

Katherine spent a lot of time determining trajectories and flight paths that airplanes and spacecrafts would need to follow to travel safely from one location to another, blast into space, orbit Earth, land on the moon, and return safe and sound.

- Use some materials to create a pathway for marbles and see if they will follow your path. If not, what changes can you make so the marbles follow your path accurately?
- Try this with dominoes, too, setting up a path of dominoes standing up. Then, topple the first one and see if the rest follow your path.

woman, and initially, many of the white men she worked with and for doubted her abilities. But she did not let that stop her! Her brilliance and confidence allowed her to do stellar work and continue to reach for the stars!

Katherine solved problems related to spacecraft trajectories (paths) for blasting off, orbiting, and landing safely. She continued to work at NASA for thirty-three years, retiring in 1986. Along the way, she received numerous awards and honorary doctorate degrees, even getting a building named after her at NASA called the Katherine G. Johnson Computational Research Facility, and receiving the Presidential Medal of Freedom from President Barack Obama on November 24, 2015.

Katherine Johnson died on February 24, 2020. She was an inspiration to all and a solid example demonstrating that hard work, determination, confidence in your abilities, and following your passions can really pay off.

Did YOU Know About . . . ?
Flight Paths

A flight path is a route that aircraft and spaceships take that helps them know where they are going as they fly. These paths can help flying machines maneuver safely through the air as they head to their destination. For example, the flight path of the *Voyager II* spaceship that launched on August 20, 1977, included flying past Jupiter and Saturn but also Uranus and Neptune before continuing beyond the solar system.

Advice for YOU

Ray Kurzweil said, "Find something that you are really passionate about! Don't try to take certain things just because it pays well or gives you a good lifestyle in terms of money. Get involved in things you are very interested in . . . If you don't have something that animates you yet, keep learning about different things and hopefully you'll find something!" He also believes that technology is constantly changing, so learn how to code because it will definitely serve you in the long run.

Ray Kurzweil

Who Is Ray Kurzweil?

Raymond Kurzweil was born on February 12, 1948, in New York City. He was a curious child who was fascinated with machines and the mechanisms that make them function. When he was five years old, his grandmother showed him a book she had written with a manual typewriter. He was intrigued by how the typewriter could change a piece of blank paper into something that looked like it came out of a book. Kurzweil studied the machine and every mechanism within it, and after that, he knew he wanted to be an inventor!

Throughout his childhood, Kurzweil collected gadgets from the neighborhood and built up an inventory that he used to construct things ranging from go-karts to a puppet theater with mechanisms that allowed the puppets to move onstage—which he built at age eight. He was convinced that if he could figure out how to put things together, he could solve any problem!

At age twelve, he discovered the computer, and even got a job using a computer to analyze programs. Then, at age fourteen, he was introduced to the artificial intelligence movement. Kurzweil was fascinated by the

relationship between computers and intelligence and was eager to learn more. Later, he applied to the Westinghouse Science Talent Search, where he highlighted a program he developed that could detect patterns in musical compositions. Not only did he win a prize at this competition, but he also got to meet US president Lyndon B. Johnson because of it, and he went on to win other competitions and showcase his work on television. From there, his practice with pattern detection grew.

After graduating from high school, Kurzweil attended Massachusetts Institute of Technology (MIT), where he completed his bachelor of science degree in computer science and literature in 1970. He continued his path of inventing, and at age twenty-five he started crafting his first major invention: a reading machine for the blind. This machine could scan a page of text and then use a voice to read the text aloud. He took this machine to the *Today* show and, shortly after, was contacted by Stevie Wonder, a famous singer-songwriter and musician who is blind. Together, Kurzweil and Wonder began Kurzweil Music Systems in 1982 and developed the Kurzweil 250, an electronic piano—also known as a synthesizer—that could use patterning to duplicate the

Science YOU Can Explore

- **With the permission of your parents or an adult, and with proper safety materials (gloves, goggles, etc.), explore a machine.**
  - ☐ Find an old machine that is not being used or that is broken.
  - ☐ Use tools, such as screwdrivers and wrenches, to take the machine apart.
  - ☐ Inspect it closely. Then ask yourself:
    - ▪ What do you notice? What types of mechanisms are within the machine?
    - ▪ What does each mechanism do and how does it help the machine work?
  - ☐ Now, try to put the machine back together, or use the materials to make something new!

sounds of other instruments. He kept at this work, developing new and improved inventions, and later expanding into artificial intelligence, which allowed him to investigate new technologies such as language launch models.

In addition to his inventions, Kurzweil used his pattern detection abilities to make predictions about technology. He wrote books and shared his thoughts with the world related to medicine, artificial intelligence, technology, exponential growth, and other scientific ideas. But even with his astounding ability to think critically, Kurzweil's path was not without challenges. Many people often disagreed with or pushed back against Kurzweil's ideas or predictions, such as scholars in the field who questioned his exponential growth prediction—that by 2029, computers will be able to pass the Turing test, meaning technology would match human intelligence. Further,

working within the field of artificial intelligence is difficult because many people are concerned that computers will replace jobs that people need to survive. However, Kurzweil believes that new jobs will emerge because of technology, and that technology will continue to make the world a better place.

Overall, Kurzweil used his interest in technology and pattern recognition to invent things that have made an impact on the lives of many, such as text-to-speech capabilities that have helped individuals with disabilities

Questions for YOU

....

1. What do you think a computer can do?
2. What do you think a computer can't do?
3. How can computers help humans in the future?

access printed materials. He has received many awards, including the National Medal for Technology and Innovation, has been inducted into the National Inventors Hall of Fame, elected to the National Academy of Engineering, and has written a number of books. He truly believes that persistence is key within the fields of science and technology, and that scientists can use their knowledge and passion to make a big effect on the world!

Did YOU Know About . . . ?
Artificial Intelligence

Artificial intelligence is a science that works to make computers mimic what humans are able to do—make computers produce pieces of art or speak in a human language, among other things. For example, the tool Chat Generative Pre-trained Transformer (ChatGPT) can use human language models to produce computer-generated responses that are similar to how a human speaks or writes. ChatGPT is powered by artificial intelligence—a computer imitating human intelligence.

Advice for YOU

....

Savage says to celebrate your failures! You can learn a lot from how you fail. It's okay if you did it wrong—learn from it and figure out how it was wrong and how you got that answer. Then, take those insights and use them to make better answers next time. If we celebrate our failures and talk about them with others, we can learn together and all be better because of it! And remember, the limit of your data science career is exponential!

Kay Savage

Who Is Kay Savage?

Kristie "Kay" Savage was born in Dearborn, Michigan. Both of her parents were engineers and shared their love of science with her. They told her at an early age that with a degree in engineering, she could do anything! When she was very young, she found her love for math because she always liked that there were right and wrong answers—they just needed to be discovered. Throughout her elementary, middle, and high school career, Savage continued taking science and math classes so she could learn as much as she could. When she graduated high school, she liked the idea of majoring in an area that would not pigeonhole her, so she took the advice of her parents, and she attended the University of Michigan with a major in engineering. Because engineering is such a broad discipline, she took a wide array of classes and learned many facets of what engineering entailed. During this time, she delved into learning more about the field she would eventually pursue: data science.

When Savage entered the work-force, she got jobs in several industries including oil, television, and advertising—each giving her a chance to use her

engineering and data science background even though she had never taken a class in data science. It was a brand-new field when she finished college, so it was not yet taught there. But the knowledge she gained about computers and math, as well as the ability to think critically about problems, allowed for her to collect data in her areas of interest and impact customers through her work. For example, she worked as a data scientist at Spotify, where she was able to study music data and then create and implement data codes to allow listeners to have personal and meaningful listening experiences.

Later, Savage moved on to a company called Fanatics, where she uses her data science skills to analyze sports data, create codes, and then offer relevant insights that impact sports teams and their fan communities. Savage also manages a team of other data scientists. She helps her team to use and

Questions for YOU

According to Savage, there are tons of data out there, and the amount will only grow—there are going to be more things to study as everyone gets smarter, and the types of questions we can answer are only going to expand. But here are two important Questions for YOU:

1. What is your most favorite thing right now?

2. How would you make it better?

If you can find those things (your favorites and ways to make them better), you can be a data scientist. That's what data scientists do and the types of problems they try to solve. Apply this to your favorite thing and you're a data scientist! That's it!

Science YOU Can Explore

You can explore anything that interests you or that you really care about!

- **Maybe you really care about your local community, parks, and playing outside. For example, you might ask, "How much fundraising might I need to do to clean up and make a new playscape in my park area?"**
  - ☐ Get some data points (size of playscape and options, effects on the environment, understanding how people and local communities might affect the environment, knowledge of where all the kinds of chemicals or pollutants are in the environment, etc.).
  - ☐ Once you have all the problems identified, start to think about potential solutions. You can make insights and predictions and put your plans into action.

hone their science skills to discover and offer important information pulled from the data and patterns they find. Then, they share their findings with others, and ultimately change lives! While Savage acknowledges that the field of science can be hard, she never lets this stop her. Some of the classes she took in college were very challenging, and even today, the problems she solves are tough—even zeroing in on a specific passion within the broad field of engineering was challenging. However, she looked for and used resources around her to help her succeed, such as taking different courses to figure out her interests, joining community and university programs to help support her education and career, visiting professor office hours to ask questions, and joining groups like the Society of Women Engineers.

Savage thoroughly enjoys pulling together her skills in the fast-changing and engaging field of data science. Her career path confirmed the idea her parents shared with her so long ago—with a degree in engineering, and the valuable skills you develop within that field, you really can craft a career path that is perfect for you! But also, she realized that finding and using resources can aid in building a community of support, which helps you overcome difficulties and allows you to continue to be successful in all you do!

Did YOU Know About . . . ?
Data Science

There are so many types of fields and domains to which data science can be applied. Scientists learn to use their computer coding skills, their statistics and data skills, and their interest in a particular domain they want to analyze.

1. **COMPUTERS** are needed to read large volumes of data. Scientists can write computer coding to allow the computer to manipulate that data and make it so a scientist does not have to read ten million rows in a database.

2. The mathematical field of **STATISTICS** helps scientists take the data and analyze it. The computer will apply statistical analysis to the large amounts of information (data), and then give insights and ideas about patterns found, or results from tests it ran.

3. Data science can be analyzed in any **DOMAIN** or area. The domain gives specifics as to the type of data a scientist will analyze. For example, in ecology, you might want to investigate how Earth moves or works. Or maybe you want to investigate customers on an app, like Uber, to figure out how fast Uber drivers drive. Perhaps it is health care, image processing, understanding brain scans, etc.?

Quote for YOU
....

Dr. Tao says, "To make the subject your own, you have to do more than just [go to] school. You have to do more at home. You have to play with those things." Dr. Tao also believes that to hone your math skills, you should play with math puzzles, make your own questions, look into and participate in math competitions, find math circles—groups that meet, pose math questions, and complete hands-on projects—near you and join them. Find ways to engage with math through exploration and locate the wonderful resources out there that can help you learn more about math!

Terence Tao

Who Is Terence Tao?

Terence Tao was born on July 17, 1975, in Adelaide, Australia. At a very young age, he discovered that he enjoyed numbers. For example, Tao would draw and shape numbers on the windows while his grandmother would wash them. He loved puzzles, math, computer games, and finding ways to make connections. His parents would often give him math tasks to keep him busy and calm. He thought he would grow up to be a shopkeeper, conducting inventory and bookkeeping, and it was not until he completed his early schooling that he realized he could pursue a job as a mathematician.

Through his early school years, he tested high on so many tests that he skipped five grades, began taking eleventh-grade math classes at age eight, and even completed college-level math courses in high school. At age sixteen, after graduating with bachelor's and master's degrees in mathematics from Flinders University of Southern Australia, Tao came to the United States and enrolled at Princeton University. He finished his PhD in mathematics and went on to complete a postdoctoral

position at the University of California, Los Angeles (UCLA), where he was later hired as a professor.

Throughout Dr. Tao's career, his talents for mathematics have influenced his field of analysis with relation to number theory and geometry in major ways. Also, he has helped connect differing fields that push overall understandings forward. His work gives him the opportunity to look for patterns within numbers, such as prime numbers, and analyze functions within those patterns. For instance, he explores questions like: Are there functions that only go up and not down, and if yes, how many functions do this, and how do they differ from one another?

While, overall, Dr. Tao faced only a few challenges, he learned that even when you have natural gifts, such as exceptional mathematical abilities, you cannot rely only on these gifts—you must also learn to study and work to continue and improve. Grit and

Questions for YOU

1. What is the biggest number?
2. How do you know?
3. How can you find out?

Science YOU Can Explore

- **Mobius Strip**
  - ☐ Take a long strip of paper. Tape it together to make a cylinder. Cut it down the center. What did you get? Why?
  - ☐ Now, take a second long strip of paper. Connect it with a twist and tape together. Cut it down the center. What did you get? Why?
- **Isoperimetric Problem**
  - ☐ Suppose you have a long rope that you tie into a loop. Which shape will allow you to enclose the biggest area with the rope? Give it a try!

hard work help support your gifts! This became very clear to him when he almost failed his qualifying exam during his doctoral program. After not studying as hard as he could have, it was apparent to him that effort and hard work are just as important as natural gifts.

Dr. Tao has received numerous awards and accolades such as the Fields Medal, the Breakthrough Prize in mathematics, the Royal Medal, a MacArthur "Genius" Fellowship, and the National Science Foundation's Alan T. Waterman Award, just to name a few. Dr. Tao continues to teach and conduct research at UCLA, while also serving on committees and mentoring the students he works with. He is often considered a leading thinker in mathematics, even being called the Mozart of mathematics.

Did YOU Know About . . . ?
Prime Numbers

You have probably heard of "prime numbers" before and you may even be able to list many of them. But in technical terms, a prime number is a number that can only be divided by itself and the number 1. For instance, consider the number 3: It can be divided by both 3 and 1, making it a prime number. What about the number 6—is that a prime number? No, because 6 can be divided by other numbers beyond itself and 1, such as 2 and 3. What about the number 1—is it a prime number? It turns out that most mathematicians say no, the number 1 is not a prime number because it is divisible by only one number, 1. As you can see, while understanding prime numbers can seem easy, there is still so much to learn about them and their characteristics, and many mathematicians are still studying prime numbers and the relationships they have with each other and other numbers.

"You want to get to the top of the cliff. But that's not what you focus on immediately. You focus on the next ledge just beyond your reach, because you need to do one clever thing to get up there. And then, once you get there, you do it again. A lot of this is rather boring and not very glamorous. But you can't jump cliffs in a single bound."

—Terence Tao

Science Activities

Formal Sciences

Activity 1:
Launch Angles

Activity 2:
Cryptography

Activity 3:
Edible Statistics

Activity 1

Launch Angles

During NASA's Project Mercury, Katherine Johnson's job was to calculate which launch angle would land a rocket where it needed to be. Now, we could do a lot of physics and geometry calculations to figure out what angle will go farthest . . . or, we can run some rocket tests!

Materials Needed

- Juice pouch
- Flexible drinking straws
- Paper
- Tape
- Scissors
- Writing implement
- Protractor (print one out online if you don't have one around the house)
- Yardstick or tape measure
- A safe spot to launch your rocket

The Experiment

1. Start by building your mini rocket. While working on this, drink your juice pouch to empty it!

2. Cut out a paper square with 5- to 10-cm-long sides.

3. Roll the paper over your straw to make a tube. It shouldn't be too tight—we need it to be able to launch easily off the straw. Tape along the seam and remove the straw from the tube.

4. Add a piece of tape to the top of the rocket (the paper tube) to seal up the opening.

5. Locate a hallway or room where your rocket can't hit anything (or anyone!), and is shielded from the wind.

6. Stick the straw into the empty juice pouch and inflate it by blowing into the straw. Place the rocket onto the straw.

7. Use the protractor to place the rocket at a 90-degree angle.

8. Give a countdown of "Three, two, one!" to make sure everyone is out of the way. Press down on the juice pouch, forcing the air out through the straw and pushing the rocket up, up, and away!

9. Measure and record the distance between your launch site and your rocket's landing place.

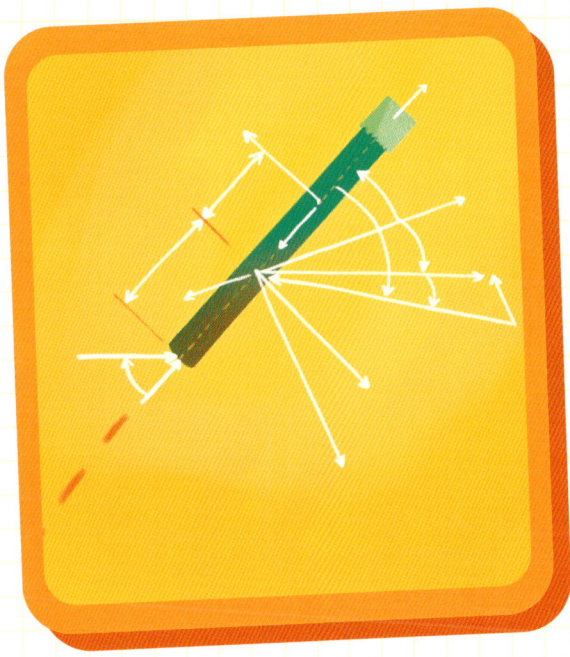

10. Repeat at different angles. Make sure you hit the juice pouch with the same force each time—if you have trouble with this, you can conduct multiple trials and take the average.

Create a data chart like this:

Launch Angle	Distance Trial 1	Distance Trial 2	Distance Trial 3	Average
90°				
60°				
45°				
30°				
0°				

- What angle sent the rocket farthest?

- A *headwind* is a wind that has the opposite general direction to a course of movement—like a current of air that blows right into the nose of an airplane. If your rocket faced a headwind, how would you change the angle?

- A *tailwind* is a wind that has the same general direction as a course of movement—like a current of air that pushes an airplane from behind. If your rocket experienced a tailwind, how would you change the angle?

- Can you adjust your experiment to add headwind and/or tailwind and test your prediction? Try it out!

Activity 2
Cryptography

A cryptographer is a person who writes (or breaks) codes for a living. The groundbreaking cryptography work that Alan Turing did in the mid-twentieth century differs greatly from the cryptography work of today. Today, most cryptographers work with computers and perform jobs related to cybersecurity and data encryption.

Despite the differences in the code breaking of today compared to Turing's time, we can trace recent advancements in cryptography to his landmark work so many years ago. His early developments in computing and artificial intelligence have helped shape the technological landscape that has continued to evolve for decades.

A code uses symbols, pictures, letters, numbers, or sounds to represent something else. A very basic code system uses letters to represent other letters. Let's try one like that!

Materials Needed

- Lined paper
- Unlined paper
- Ruler
- Pencil, pen, or markers
- Paper clips
- Scissors
- Round items to trace
- Compass (optional)
- Metal paper fastener (optional)
- Computer (optional)

Crack the Code #1

YJYL_RSPGLE'Q_UMPI_

APYAIGlE_RFC_"CLGEKY"_AMBC_FCJNCB

CLB_RFC_QCAMLB_UMPJB_UYP.

Transfer this code to a piece of paper or into a word processing application on your computer and look for clues to help you crack the code and find the hidden message. Remember, each letter represents another letter. Here are some hints to help you get started:

- There is a simple pattern in this code.
- The letter *U* represents the letter *W*.
- The letter *P* represents the letter *R*.

Did you find the hidden message in the code? Did you see the pattern in this code? Did you start with the small words or the larger words? Were there any additional clues that helped you crack this code? Was there a point where you felt like you weren't making any progress? If so, how did you proceed? And why did you keep going?

** Answer on p. 36 **

Crack the Code #2

Here's a code that uses numbers to represent letters of the alphabet. Try transferring this code to a piece of paper or your computer and see if you can find the hidden message.

3 9 16 8 5 18 - 23 8 5 5 12 19 - 1 18 5 -

20 15 15 12 19 - 21 19 5 4 - 20 15

__ __ __ __ __ __   __   __ __ __ __ __   __ __ __ __

__ __ __ __ __ __   __   __ __ __ __ __

3 18 5 1 20 5 - 15 18 - 4 5 3 9 16 8 5 18 -

3 15 4 5 4 - 13 5 19 19 1 7 5 19

__ __ __ __ __ __ __ __   __   __ __   __ __ __ __ __ __ __ __ __ __

__ __ __ __ __   __ __ __ __ __ __ __ __ __ __

Some codes use identifiable patterns, like the codes we just cracked, and some codes don't. This is true when we're developing codes for fun, like this, and it's also true for professional cryptography and data encryption.

What are some benefits to developing a code with a discernable pattern versus a code without a pattern? Do you think there are different purposes for each?

** Answer on p. 36 **

Crack the Code #1: Solution

Y J Y L _ R S P G L E ' Q _ U M P I _

A P Y A I G L E _ R F C _ "C L G E K Y"

A L A N _ T U R I N G ' S _ W O R K _ C R A C K I N G _ T H E _ "E N I G M A"

_ A M B C _ F C J N C B

C L B _ R F C _ Q C A M L B _ U M P J B _ U Y P.

C O D E _ H E L P E D _ E N D _ T H E _ S E C O N D _ W O R L D _ W A R.

What's the pattern? Each letter represents the letter two spots before it in the alphabet. A = C, B = D, C = E, etc. Did you notice punctuation in the code? Did that help?

Crack the Code #2: Solution

3 9 16 8 5 18 _ 23 8 5 5 12 19 _ 1 18 5 _

20 15 15 12 19 _ 21 19 5 4 _ 20 15

C I P H E R _ W H E E L S _ A R E _
T O O L S _ U S E D _ T O

3 18 5 1 20 5 _15 18_ 4 5 3 9 16 8 5 18 _

3 15 4 5 4 _ 13 5 19 19 1 7 5 19

C R E A T E _ O R _ D E C I P H E R _
C O D E D _ M E S S A G E S.

What's the pattern? 1 = A, 2 = B, 3 = C, etc.

Activity 3
Edible Statistics

A data scientist's job is to find patterns in large sets of information. Most of the time, we can't parse through *all* the data that exists without using a lot of computing and human resources. It's much more efficient to take a smaller sample of that data—the smaller the sample, the easier it is to parse, but the less representative it is of the greater set. Let's take a look at how small we can make our sample size using a tasty example.

Materials Needed

- Bag of Skittles, M&M'S, trail mix, Chex Mix, Lucky Charms, or any other mixed collection of snacks

- Pen or pencil and paper
- Calculator (optional)

The Setup

Draw out a table like the one below. The first column should be labeled *n*, which is our sample size—or, how many. Then, all other columns are the different categories of snacks. If you're using a bag of trail mix, your table might look like this:

n	Raisins	Peanuts	Chocolate	Cereal	Pretzels

The Hypothesis

Just for fun, make an educated guess on how many of each item you think will be included in your sample.

The Data Collection

The number *n* is our sample size, which means all the snack items within our sample. Open your bag of snacks and take a sample of only one snack item, *n* = 1. *No peeking!* Write down a number under the category your sample belongs to. For example, if you pulled one raisin, mark "1" under "Raisins" in your table.

n	Raisins	Peanuts	Chocolate	Cereal	Pretzels
1	1				

After tallying up each of your snack items, you can calculate the percentage of raisins in your sample of 1 using the formula

$$100 \times \frac{\text{\# of raisins}}{n}$$

The probability here is 100 percent. Write that down, too!

n	Raisins	Peanuts	Chocolate	Cereal	Pretzels
1	1 100%				

The table here now includes our sample of 1. This isn't a very representative sample—we can't just assume that since we got one raisin, the whole bag is filled with raisins.

Let's try a larger sample, $n = 5$. Mark down how many there are and calculate the percentages. For example, if you pulled one raisin, one peanut, and three chocolates, your table would look like this:

n	Raisins	Peanuts	Chocolate	Cereal	Pretzels
1	1 100%				
5	1 20%	1 20%	3 60%		

We got 20 percent by doing $100 \times \frac{1}{5}$, and 60 percent from $100 \times \frac{3}{5}$.

Keep going for larger and larger sample sizes! (Hint: Calculating percentages is much easier if you pull multiples of five or ten.) Here's an example of a table:

n	Raisins	Peanuts	Chocolate	Cereal	Pretzels
1	1 100%				
5	2 40%	1 20%	1 20%		1 20%
10	2 20%	5 50%	2 20%	1 10%	
20	5 25%	4 20%	7 35%	4 20%	
25	4 16%	4 16%	10 40%	5 20%	2 8%

As n gets larger, you'll notice that the percentages start to settle into a pattern. For this data set, for example, we can surmise that the person who made the trail mix really likes chocolate or might be running out of pretzels.

In each of your trials, what are some things you noticed about the percentages?

The Analysis

There are many ways data can be distributed, and we don't have to count all the data points to make some educated guesses about where they lie. Take a look at the graphs of our trail mix data set for $n = 25$, $n = 100$, and $n = 10,000$:

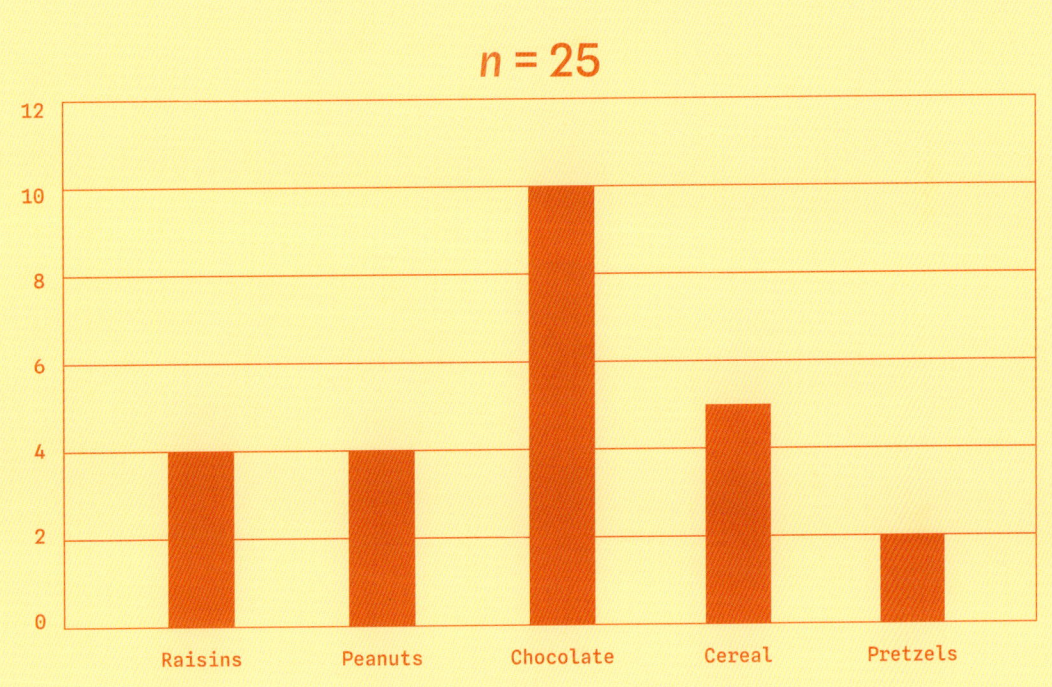

n = 100
45
40
35
30
25
20
15
10
5
0
Raisins
Peanuts
Chocolate
Cereal
Pretzels

n = 10,000
4,500
4,000
3,500
3,000
2,500
2,000
1,500
1000
500
0
Raisins
Peanuts
Chocolate
Cereal
Pretzels

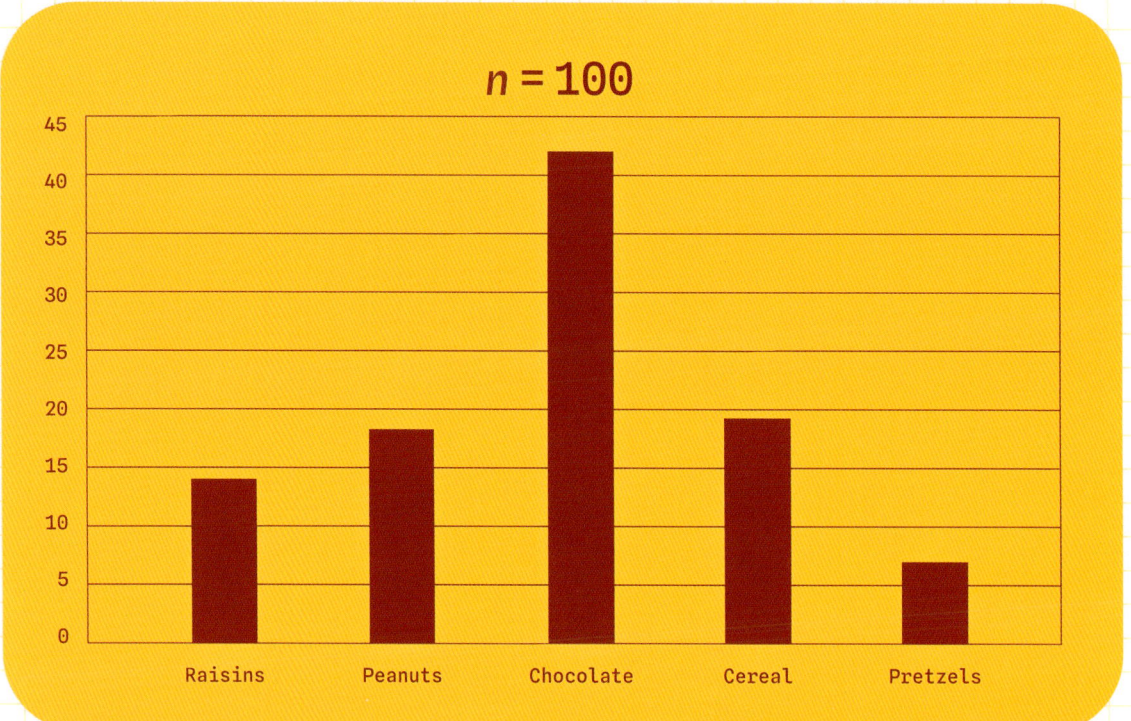

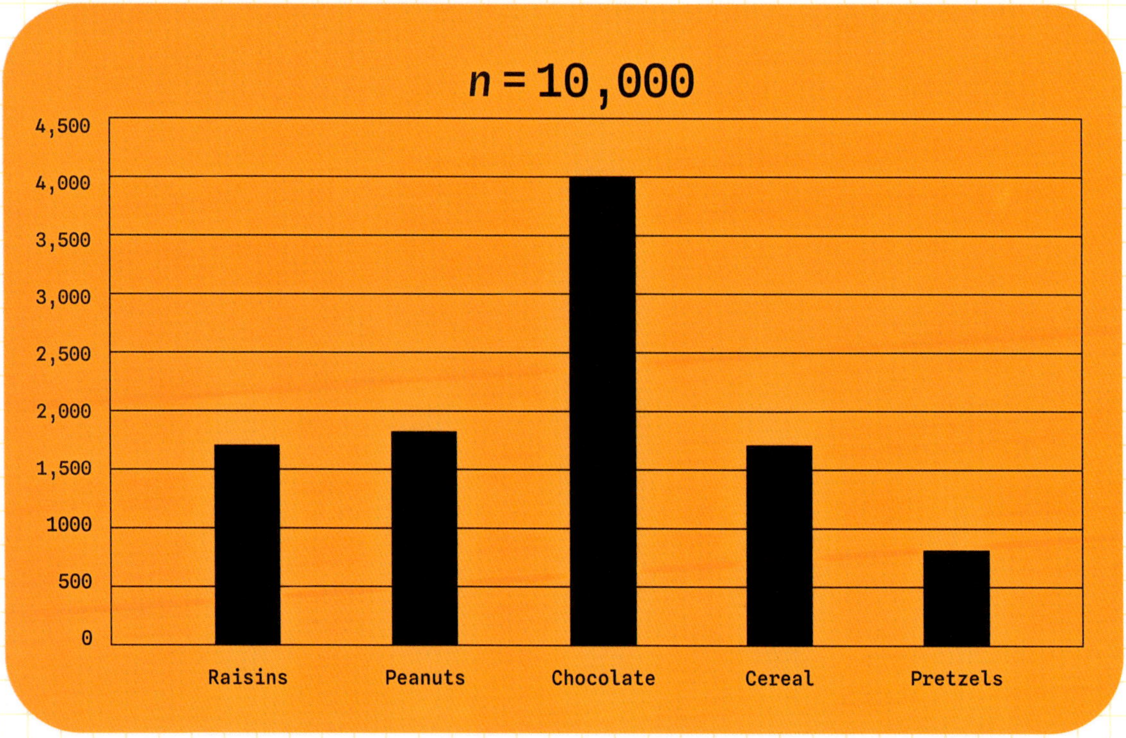

Physical Sciences

Physics

"Why do things fall down instead of up?"

Chemistry

"Why does soap make bubbles?"

Atmospheric Sciences

"What is ozone?"

Climatology

"What is the difference between climate and weather?"

Meteorology

"What is the atmosphere made of?"

Mechanics

"What is torque and how does it work?"

Geology

"What causes earthquakes?"

Astronomy

"What makes stars twinkle?"

Physical Sciences

Natural sciences involve studying the natural and physical world. When considering the physical world, this includes such disciplines as physics and chemistry, as well as Earth sciences such as astronomy, atmospheric sciences, climatology, geology, and meteorology.

Mariah Gladstone
1993–Present

Environmental and Food Systems Scientist

Michael E. Mann
1965–Present

Climatologist and Geophysicist

Pedro Sanchez
1940–Present

Environmental and Soil Scientist

Ellen Ochoa
1958–Present

Astronaut, Engineer, and Physical Scientist

Chien-Shiung Wu
1912–1997

Physicist

Advice for YOU

....

Gladstone says that everybody should learn about the plants in their neighborhood and which plants are safe to eat and how you can make them safe to eat. Also, don't be afraid to ask questions! Then go out and start learning to find answers to those questions.

Mariah Gladstone

Who Is Mariah Gladstone?

Mariah Gladstone was born on December 1, 1993, in Kalispell, Montana. She is Blackfeet on her father's side and Cherokee on her mother's side. As a young child, Gladstone spent much of her childhood outdoors exploring the woods, playing in gardens, and learning to cook from her mother and others in her family. At the young age of four years old, she was tasked with planting seeds and growing food in her own garden, which she would create and write recipes for, and then share the food she grew with her family members. In addition to loving nature, she also loved math, and her mother would often allow her to experiment in the kitchen, providing hands-on and meaningful math and cooking experiences that laid the foundation for so many of Gladstone's passions. From these experiences, she realized she wanted to pursue a career path that would help her make a difference in the environment. She went to Columbia University, from which she graduated with a degree in environmental engineering.

After completing her bachelor's degree at Columbia, she decided she wanted to continue her work exploring the growth and use of Indigenous foods that grew on the North American continent as well as the food systems related to them. It was clear to her that the field of food systems was strongly rooted in science, combining botany, ecology, land management, and nutrition sciences—an intersection that comes together to help feed humans but also takes care of the world around us. Thus, in 2016, she began her work on Indigikitchen. Here, she creates cooking videos and other material to expose viewers to recipes utilizing Indigenous and native foods as a way to spread knowledge about supporting healthy bodies and families, as well as the ecosystems that support the growth of these Indigenous foods.

Gladstone returned to school to complete her master's degree in environmental science coupled with natural and

Science YOU Can Explore

- **What is your plant knowledge? What berries do you know? Which trees or weeds?**
  - ☐ With an adult, go out and learn to identify plants in your community and get to know them. Spend days exploring and learning about plants, trees, and bushes in your backyard, neighborhood, parks, and town, and feel more connected to them.
    - Plants are your relatives. Know their stories. Which are dangerous? Helpful? Edible? Edible after some type of process that helps make them safe?
    - Share that knowledge with your friends so other kids know that information, too.
    - If you are familiar with one plant and your friends aren't, share that information so you all can know that plant. You can teach each other.

human systems at State University of New York.

Throughout her career, Gladstone has held challenging positions, such as being a steel gang roadmaster. But one of her biggest challenges is being her own boss. As the owner of Indigikitchen, her own science-related business, she has had to find the most effective ways to manage her time, take on projects that best align with her goals and interests, and utilize the resources she has while locating others she needs. Further, she is very interested in continuing to explore ways we can better take care of our bodies and the places our foods come from.

Despite the difficulties she has faced, Gladstone has found a way to incorporate her scientific knowledge, her culture and heritage, and her love of food and cooking into a solid career that allows her to help humans as well as the ecosystems with which humans coexist. Currently, she travels all around sharing her knowledge about Indigenous foods, recipes, systems, and how to implement them for stronger health and communities.

Questions for YOU

1. How can we take care of our bodies better so we have the energy to do what we want to do?

2. How can we work to take in more nutritious foods that our bodies need and cook things that are healthy?

3. How do we take care of the places our foods come from so we make sure our children and our children's children, and beyond, not only have healthy foods but also can get food from those places in the future?

The Three Sisters

Some plants like to grow alongside others. When planted with other plants, they thrive and help each other survive. This is true of what Gladstone refers to as the "Three Sisters":

Among Indigenous communities, the Three Sisters are three plants that, when planted near each other, help the others flourish. They include (1) corn, which stands tall and straight as it grows; (2) climbing beans, which need something to hold on to as they grow but that also pull nitrogen from the ground that could harm the other plants; and (3) squash, whose leaves spread out along the ground providing shade to keep the soil cool.

As Gladstone explained, these plants support each other physically, but also make up a complete protein when eaten together.

Gladstone also described that while there are some plants that act as companion plants when grown together, some plants do not thrive well together at all. For example, she noted that there is sometimes a fourth sister—sunflower. If sunflower is planted right next to the Three Sisters, it could be harmful to their growth. But, if sunflower is planted off to the side, bordering the Three Sisters, this plant can help keep weeds from encroaching into the Three Sisters' space, and the bright, beautiful petals of a sunflower can attract pollinators, like bees, to sip from the plants' flowers. Gladstone noted that the Three Sisters, in cultures that grow them, serve as a metaphor for how people can support each other, even if they have different characteristics.

Quote for YOU

....

Dr. Mann says: "The choices we are making today, at this very moment, are going to determine the sort of planet that [children] will inherit as they grow up. It is important to be engaged NOW, because we can't afford to wait decades from now. The decisions we are making today, especially true in the climate arena . . . are going to impact the planet for decades and centuries to come. It is never too early to become involved in any way . . . Children can become a part of the conversation, talk to adults, convey the importance of the health of our planet. If we don't act now, it will soon become too late to prevent catastrophic consequences."

Michael E. Mann

Who Is Michael E. Mann?

Michael E. Mann was born on December 28, 1965, in Amherst, Massachusetts. His father was a professor of mathematics and his mother was a teacher and homemaker. Thanks to his exposure to the academic community, he fell in love with science at an early age and was fascinated with nature and understanding how the natural world works. Even as a young child, he would ask questions about nature and was particularly interested in natural disasters such as hurricanes, tidal waves, tsunamis, and other natural phenomena.

Throughout his elementary, middle, and high school years, he was captivated by science and was especially inspired by Carl Sagan and documentaries he would watch about science, and he strived to do similar work. During his time in high school, he became intrigued with computer programming, mainly solving problems and creating computer programs to help. After graduating high school, he realized that in college he could combine science with computer programming, so he double majored in applied mathematics and physics at the University of California, Berkeley, and later went on to Yale University to complete his PhD in geology and geophysics.

Questions for YOU

1. How is climate change impacting our daily lives?
2. How is climate change impacting our lives now in ways we may not realize?
3. How can we contend with all of these climate-related challenges?

During his time at Yale, Mann worked with an influential professor and began working on models related to Earth and climate. Based on his and his colleagues' work, he published a jarring finding called the "Hockey Stick" graph, which revealed how quickly Earth was warming due to climate change. Although his work was foundational in arguing for the importance of addressing climate change, it also put him in the crossfire of political battles that challenged the science because it could hurt business-related profits. This was a major obstacle for him, as his work was questioned and others tried to discredit his findings.

However, he did not let this stop him. Instead, he used this as fuel to continue his research and work as an advocate to fight for addressing climate change—even taking some individuals to court for defamation, and winning! He travels the world and speaks about changes society will need to make to address climate change before it is too late.

Science YOU Can Explore

- **Go out and do some simple research on climate change.**
  - ☐ What is climate change?
  - ☐ What causes climate change or makes it worse?
  - ☐ What are some examples of catastrophes caused by climate change?
  - ☐ What are some steps we can take to minimize the impact of climate change?
- **Explore some of the youth advocates around the world and see what they are doing about the major concern of climate change. Here are some youth advocates you can learn more about:**
  - ☐ Artemisa Xakriabá
  - ☐ Autumn Peltier
  - ☐ Greta Thunberg
  - ☐ Leah Namugerwa
  - ☐ Qiyun Woo
  - ☐ Xiuhtezcatl Martinez
  - ☐ Zanagee Artis
- **Now, think about what you can do in your home, school, and community that can contribute to combating climate change around you!**

Dr. Mann continues to use his understanding of physics, chemistry, and biology to describe how our planetary system functions; to learn how to address questions about climate change; to figure out what steps we must take to prevent further change; as well as to understand the catastrophic effects climate change can have on Earth. He also works to explore how climate change affects extreme weather events such as flooding, heat, and wildfires and what we can do to solve these problems, too.

Dr. Mann and his body of work have gleaned numerous honors, such as him being inducted into the US National Academy of Sciences, being named Humanist of the Year for 2023, contributing to the work that won the 2007 Nobel Peace Prize, and being known for working as a staunch advocate to fight climate change.

Did YOU Know About . . . ?
Climate Change

Climate includes the weather in a certain area that is the average based on a long period of time, which can include that location's average temperature, rain or snow fall, wind, clouds, sunshine, and so on. When considering climate over a long period of time, it becomes possible to figure out what is considered typical. However, changes in Earth's atmosphere, especially related to human behaviors, can cause the climate to change. These changes can have major and lasting effects on the climate and can lead to major catastrophes such as more wildfires, stronger hurricanes, and hotter temperatures. By understanding how our actions lead to climate change, we can begin to work to change our behaviors to protect the planet we live on!

Areas of Scientific Interest

Space, Engineering, and Physical Sciences

Advice for YOU

Dr. Ochoa says that science is about curiosity! By applying the process of science (asking questions and going about answering them), you can learn new things. That is the power and excitement of science!

Dr. Ochoa also thinks it's important to find your supporters! Find those who know you and know what you are capable of bringing to the table (hard work, perseverance, willingness to ask questions, dedication to an organization's mission, etc.).

Ellen Ochoa

Who Is Ellen Ochoa?

Ellen Ochoa was born on May 10, 1958, in Los Angeles, California, and she and her four siblings grew up in La Mesa, California. As a child, she was more interested in music than science and began playing the flute, which she continued to play throughout her childhood and adulthood! Even though she loved music, she also enjoyed math, and she took all the math classes her high school offered.

When Ellen graduated high school and went to college, she was unsure of what to major in, so she decided to explore subjects related to the math classes she had completed previously. First, she visited a professor of engineering who dismissed her because he had no experience with women being engineers. Fortunately, this did not stop her interest in exploring the fields of science and math, and she went on to meet a professor of physics and decided to pursue this as a possible career path. She studied physics as an undergraduate and graduated from San Diego State University with her bachelor's degree after completing her coursework and participating in fellowships. She then completed a master's and a PhD in engineering from Stanford University.

During her graduate-level studies, Dr. Ochoa focused on optics and worked with her advisers to develop a process used within the field of optical systems—they even received patents for their

Questions for YOU

1. What type of person do you picture when someone says the word "scientist"?

2. Now, can you picture *yourself* as a scientist? Try it out, because **YOU** can be one, too!

inventions. In her first job after graduate school, Dr. Ochoa continued developing inventions that also received patents. She faced challenges along the way, such as being a woman in the math and science fields, which were often male-dominated spaces where many people did not believe women belonged. For example, when Dr. Ochoa was in college, no American woman had ever been to space. But while she was in graduate school, Sally Ride became the first American woman in space. Ride and other women astronauts inspired Dr. Ochoa to consider applying to the astronaut program once she finished her doctorate degree. At first, she was not selected, and she realized she needed some operational experience as well as continued work experience. So, she got a pilot's license and took a research job at the National Aeronautics and Space Administration—NASA! With these improved skills, she reapplied to the astronaut program and was accepted.

Dr. Ochoa's brilliance, hard work, knowledge, and experience allowed her to not only train to become an astronaut—as the first woman of Hispanic descent to travel to space—but also to fly on four missions; study a variety of variables in space including ultraviolet light and its interactions with Earth's atmosphere and ozone layers; help assemble the International Space Station (ISS); operate a robotic arm in space; serve on committees that selected

Science YOU Can Explore

- **Do some internet research about space!**
  - ☐ What do astronauts have to do to survive in space?
  - ☐ Have any animals been to space?
  - ☐ What is the ozone layer?
  - ☐ How long does it take to orbit planet Earth?
  - ☐ Now, explore some of your own questions about space!

multinational crews; and serve as a director of the Johnson Space Center. Dr. Ochoa even got a chance to play her flute in space! Her hard work and determination inspires young scientists all over the world, and illustrates that you can become whatever you put your mind to, even an astronaut conducting science in space!

Did YOU Know About . . . ?

The International Space Station

NASA, or the National Aeronautics and Space Administration, is a government agency that conducts scientific and technological research related to air and space for the United States. One specific resource NASA is involved with is the International Space Station (ISS). This is a huge spacecraft that serves as a scientific laboratory where scientists and astronauts from all over the world can conduct unique science experiments that you would not be able to do on Earth. The ISS orbits Earth and moves 17,500 miles per hour, meaning it moves completely around Earth every 90 minutes.

People have been living in space since November 2000. Usually, there are seven people at a time on the ISS, and there's a Chinese space station that, generally, has three people. Most of the crew members spend about six months in space, then a new crew comes up and they switch out. Some people have spent about one year at a time on the ISS. Now, thanks to the work of scientists like Dr. Ochoa, there are scientists doing science in space twenty-four hours a day, seven days a week.

Want to learn more about NASA? Check out some of the resources on its website: NASA.gov/stem.

Areas of Scientific Interest
....

Environmental and Soil Sciences

Quote for YOU
....

Dr. Sanchez says, "You never know, but something that you really like, such as red dust washing from your skin and down the drain, can turn into a lifelong career in science." He also suggests that you be curious about our world, where we are, and the things around you, and if there's something you like, or a scientific interest, go out and explore it!

Pedro Sanchez

Who Is Pedro Sanchez?

Pedro Sanchez was born on October 7, 1940, outside Havana, Cuba. When he was younger, his family owned and ran a farm, so as a child, he learned about crops and agriculture from his father, a master farmer. For example, at a young age, he learned about how crops grow and how to handle farming tools carefully. But for young Sanchez, his interest in science was really piqued each time he stood in the shower after a long day of farming and watched the red soil flow off his body and down the drain. This captivated him and sowed his interest in learning more about soil.

When his family was forced to flee Cuba for the United States, he decided to continue his studies, with the mission of dedicating his life to solving problems related to soils found in tropical regions. He went to college at Cornell University and completed his bachelor's and master's degrees in soil science, and his PhD in tropical soils, because he knew that tropical soils were found in many places where the poorest people in the world live.

Early in his academic career, and despite obstacles he faced, Pedro visited places all over the world to learn more about soils. The vast amount of knowledge and expertise he gained gave him a chance to teach farmers in places such as Brazil, Peru, and the Philippines how to grow important crops, such as rice, to feed themselves, their families, and their communities. Across his career, his work took him to

Questions for YOU

1. What makes you curious?
2. Where can you find soil?
3. What kind of soils are on Mars and other planets?

other places, such as Kenya, when he assumed the role of the chief executive officer of the World Agroforestry Center (ICRAF), and later to Nairobi and at least twenty other countries throughout Africa to help solve soil infertility issues and, ultimately, grow food. In addition to losing everything when his family had to flee Cuba to the United States, he also faced other challenges, such as having to work very hard to ensure that his citizen documents were in order so he could travel and conduct his research, and surviving two bouts of cancer and a bad case of malaria. However, he was able to overcome these difficulties to carry out his lifelong calling of assisting people all over the world.

Dr. Sanchez has dedicated his life to helping people learn how to grow food as a way to combat worldwide starvation and food insecurity, as well as solving problems related to soil infertility and the management of soils. He has been a professor and researcher at colleges such as North Carolina State University, University of Florida, and Columbia University and has earned a large number of accolades and awards, such as

Science YOU Can Explore

....

- Soil is a complicated thing. It's alive because so many microbes and microorganisms live in it. So, go outside and mix water into different types of soil. Watch what happens.
- Look at the different sizes of particles in soils and dirt such as sand, clay, etc.
- Try this experiment:
  - ☐ Plant some seeds, like corn, in different types of soil. See what happens when you plant corn in sand, silt, and clay.
    - What happens when you add fertilizer?
    - Jot down your observations in a notebook over time.

receiving four honorary degrees, writing a groundbreaking book on soils, which is still considered the bible of soil science, becoming an honorary chief to two tribes in Africa, serving on advisory boards for US president Joe Biden, and more. He retired from the University of Florida in May 2022. But, as Dr. Sanchez stated, he believes the best accolade is when a farmer thanks him and tells him that because of Dr. Sanchez's work, that farmer and his family no longer must go hungry!

Did YOU Know About . . . ?
Soil

At the very top part of Earth's crust is where you will find soil. This thin layer is made up of rock, minerals, decaying animals, and materials from plants. Plants require soil to grow because they pull minerals and water from it. Further, soil can range from being sandy to loamy to clay-like. In places where temperatures are high but the soil still receives high amounts of rain, tropical soils can be found. Research shows that tropical soils are not the best for growing plants because of their high acidity, which makes it difficult to grow crops and food.

Quote for YOU
....

Dr. Chien-Shiung Wu said, "I have always felt that physics, and probably in other endeavors, too, you must have total commitment. It is not just a job. It is a part of life"—so if you find something you are passionate about, do your best to give it your all.

Another piece of advice that was important to Dr. Wu came from her father when she was very young. He stated, "Ignore obstacles. Just put your head down and keep walking forward"—and that was a piece of advice she took to heart—and you should, too!

Chien-Shiung Wu

Who Is Chien-Shiung Wu?

Chien-Shiung Wu was born on May 31, 1912, in Liuhe, China. In her language, Chien-Shiung means "courageous hero," which was the name given to her by her father, and it set the stage for her entire life. When she was young, many people in China believed that girls could not accomplish the same things boys could and, instead, expected girls to be homemakers. Because of this opinion, they also did not believe in sending girls to school. Fortunately, Chien-Shiung's parents felt differently. Her father, who was an engineer, quit his job and opened a school that taught girls important skills such as math and reading. Chien-Shiung was among the pupils at her father's school, and her

mother taught there, too. After several years of teaching Chien-Shiung, her parents realized she needed more—they had instructed her as much as they could.

Because there were no other schools nearby that taught girls, she was sent fifty miles away to a boarding school where she continued her education. Although she had the choice of either teacher-training or a more rigorous program, she selected teacher-training because it was a free program, but she

borrowed textbooks and materials from her classmates who were in the other program. Ultimately, she taught herself the more academic content, such as mathematics, physics, and chemistry, and excelled in her studies, graduating at the top of her class! Her talent was recognized, and she was selected to go to National Central University in Nanjing, China, to continue her studies.

After Chien-Shiung graduated from National Central University in 1934, she decided to pursue graduate studies and left China for the United States, where she registered at the University of California, Berkeley, and completed a PhD in physics. She continued to apply her incredible work ethic to her study of physics and worked as a research assistant. Shortly after, she and her husband moved to New York and she took a teaching position at Smith College, and then a professorship at Princeton as the first female instructor at the

Science YOU Can Explore

- **Go out and watch physics in motion! Set up a ramp out of planks and roll a ball down the ramp.**
  - ☐ What happens when you lift the ramp so the incline is very steep? What about when it isn't steep—how does that change how the ball rolls?
  - ☐ Place different objects at the bottom of the ramp (another ball, a wooden block, a shoe, etc.) and roll the ball down the ramp. What happens to the object? What happens to the ball?
  - ☐ Find other objects and see what happens when they interact with each other. Then make small changes and note the differences in their interactions.

school. These positions did not come without difficulties—she often faced discrimination for being Asian and also for being a woman in science. She was sometimes passed over for prestigious awards and positions because of her identity. But she pushed on, studying physics, teaching, and conducting top-notch research, especially on a phenomenon called beta decay, which she became an expert on and designed an experiment to prove its existence—even when the physicist who identified the phenomenon could not prove it himself. Word of her expertise spread throughout the academic community, and she was asked to join a team at Columbia University to work on a confidential task called the Manhattan Project, which gave hand-selected scientists the task of building an atomic bomb to help the United States during World War II.

After the war was over, she continued teaching and conducting research at Columbia and her reputation grew as more scientists reached out to her to help them figure out problems they could not solve. Across her life, in addition to conducting incredible experiments and becoming a leading physicist, she spent time amplifying her voice about issues tied to war and gender equality for women in sciences. Dr. Wu retired from Columbia University in 1981 and died on February 16, 1997. She had a long and impactful career, during which she achieved a number of accomplishments and earned awards and accolades, such as being

Question for YOU

Science can often help people far and wide, and can help a scientist serve their country. Can you think of other scientists who used their knowledge to help their country? In what ways did they help?

inducted into the National Academy of Sciences, as well as being a recipient of the National Medal of Science given to her by US president Gerald Ford, and the Wolf Prize in Physics in 1978, just to name a few. Further, she was hailed as the "Queen of Physics" by *Newsweek* magazine and "the First Lady of Physics Research" by *Smithsonian* magazine.

Did YOU Know About . . . ?
Physics

Physics includes the study of matter and energy and how the two interact. Matter is anything that takes up space and has mass, whereas energy is the capability for something to do work. Energy can take many forms, ranging from light and heat, to electrical, mechanical, chemical, even nuclear! So, when a large ball rolls down a ramp and hits another ball, the reaction that happens is physics in action!

Science Activities

Physical Sciences

Activity 1:
Liquefaction

Activity 2:
Soil pH and Kale

Activity 3:
Collaborate in Space

Activity 1

Liquefaction

One type of natural disaster that occurs on Earth is an earthquake. Earth's top-most layer of the mantle is made of small pieces of Earth's crust (composed of slabs of rock and called tectonic plates) instead of a solid sheet. The pieces move around slowly, constantly passing each other and occasionally bumping into one another. The tectonic plates' edges, or faults, are rough and get stuck together. When they become unstuck, an earthquake occurs and sends vibrations through the soil.

Rising water levels caused by climate change can contribute to the extent of an earthquake's damage. The rising waters can increase the possibility of a phenomenon called "liquefaction." Liquefaction occurs when the loosely packed, water-logged sediments at or near Earth's surface lose their strength due to the strong shaking of the ground. The water-filled spaces between the grains allow the sediment to flow like a liquid!

In this activity, you are going to explore the effects of liquefaction and how it can change the Earth!

Materials Needed

- Deep plastic container (as large as you like)
- Sand, enough to fill the plastic container
- Large cup filled with water
- Spoon
- A light object (Ping-Pong ball or egg)
- A heavy object (wooden block or weight)
- Paper towels for cleaning up

The Experiment

1. Fill the container nearly to the top with sand.

2. Add water slowly to the sand, but make sure not to add more water than sand. Give the container a few shakes. If it begins to liquefy, you have added enough water. It will look like wet sand with some puddling on top.

3. Smooth and press the sand down with the back of the spoon to push the water underneath the sand.

4. Bury the lightweight object under the sand on one side of the container. Set the heavier object on top of the sand on the opposite side.

5. Place the container on a countertop. Using your hands, pat the countertop next to the sand container to mimic an earthquake. You can also gently bang the entire container on the countertop.

6. Observe the objects shift and move—that's liquefaction.

If you feel like doing more:

1. Increase the strength, duration, or frequency of the shaking of the bin. How does that affect the way the objects move?

2. Try the experiment with different amounts of water.

3. Collect different soil samples from around your home with different granular sizes. Try the experiment with the same level of water and the different soil samples.

Soil pH and Kale

Understanding the science of soil can help you to be a more successful gardener of plants, fruits, and vegetables. One of soil's markers is its pH level. Some plants prefer soil that is acidic, whereas others like an alkaline environment.

Materials Needed

- A location with dirt, or your garden
- 2 tablespoons
- 2 small bowls
- Distilled white vinegar
- Baking soda

The Experiment: Soil Testing

1. Dig below the dirt's surface, at least 2 inches deep, to collect a soil sample.

2. Scoop 3 tablespoons of soil into each bowl.

3. To the first bowl, add 6 tablespoons distilled white vinegar.

4. To the second bowl, add 2 tablespoons water and 6 tablespoons baking soda.

5. What was the reaction in each bowl?

If soil reacts to the baking soda, it is considered acidic. If it reacts to the vinegar, it is alkaline. If there is no reaction, it may be neutral or inconclusive. Test kits are the most accurate way of testing the makeup of soil.

Considering Kale

Kale is a leafy plant that belongs to the plant family known as Brassicaceae. Other plants in this family are broccoli, Brussels sprouts, cabbage, cauliflower, collard greens, and kohlrabi. While kale is a robust plant that can grow in most conditions, it prefers soil with a more neutral pH level. When the environment is right, it has a fast growth rate and grows in about three months. In addition to growing easily, kale contains health-promoting phytochemicals and has nutrients such as calcium and vitamin K, which help build healthy, strong bones.

Materials Needed

- Soil with a neutral pH
- Garden spade
- Kale seeds
- Water
- Compost

The Experiment: Grow Kale

In this activity, you are going to grow your own food (or explore how food is grown) and use the yield to create a delicious and nutritious snack.

1. Find a good planting location with full sun.

2. Begin your garden about 2 weeks before the last frost—you can check *The Old Farmer's Almanac* (www.almanac.com) to learn when the last frost will occur in your area. Plant the kale seeds ¼-inch deep and about 12 inches apart.

3. Nitrogen, often found in compost, among other places, helps leaves grow. Add a few inches of compost to increase the health of your plants.

Materials Needed

- Baking sheet
- Parchment paper
- 1 bunch of kale picked at the stem from your garden, or the grocery store
- Large bowl
- 1½ teaspoons oil

- ⅛ teaspoon salt
- ⅛ teaspoon ground black pepper, garlic salt, or preferred spices
- Mixing spoon

The Experiment: Eat Kale

1. Preheat the oven to 350 degrees. Line a baking sheet with parchment paper and set aside.

2. Rinse the kale leaves to remove any dirt (adding a bit of vinegar and baking soda to your wash helps clean produce well).

3. Rip the leaves into small, 1-inch pieces and place the stems off to the side. The stems are ideal for soups or compost.

4. In a large bowl, using your clean hands, mix the kale leaf pieces with the oil, salt, and seasonings until they are all coated. Once coated, place the kale pieces on the prepared baking sheet, spreading them out.

5. Bake the kale chips for 10 to 15 minutes, carefully flipping them over halfway through the baking time.

6. Dig in—you just made your very own kale chips!

If you feel like doing more:

1. Try the soil experiment and compare it with an official soil test kit to see how your experiment matched up.

2. Try growing kale in different locations; indoors in a container garden, in a location with partial shade, or with different soils to see how the leaves grow differently based on nutrients it gathers from the earth.

3. Try baking kale chips with different flavors to see how spices change the way kale tastes.

4. Kale can be eaten cooked or fresh! Explore different ways to enjoy the healthy plant.

Collaborate in Space

Have you ever tried to work with a partner, or several partners, to get something done? Was it easy? Was it hard? Was if fun? Imagine working with a partner if you were in separate rooms, countries, or not even on Earth! Then you would be tackling some of the same job functions as Ellen Ochoa, former astronaut and director of the Johnson Space Center.

When Ellen was an astronaut, it was part of her mission to perform research experiments in space. To get the best results, she might work with fellow astronauts in space and on the ground, possibly in different places around the world! To be sure they're all working together efficiently and putting the right materials together in good order, they would have to communicate clearly.

As the director of the Johnson Space Center, Ellen oversaw many projects with team members working all over the world and in space. Part of her job was to give everyone clear instructions, feedback, and reactions to how things were going, and to teach all teams to be able to do that, too. That's clear and effective communication!

In this activity, you will be part of a team working together to build a structure necessary for your next experiment to be carried out on the ISS, or International Space Station.

The Setup

1. Within your group, figure out whose birthday is closest to today's date. That person will be the mission director. Everyone else in the group will be astronauts hard at work on the ISS.

2. To conduct this next experiment, you need to assemble a structure. Everyone on the ISS has the component pieces and, on Earth, the mission director has the instructions as well as the pieces to use for building.

3. Due to the nature of materials storage, all the astronauts are in different compartments within the ISS. They cannot see what their fellow astronauts are building and they cannot see the set of instructions the mission director is using.

4. To model this, all team members will sit around a table with a manila folder, or some other barrier, placed in front of them to be sure no one can see the team members' materials. If this isn't available, sit in the space so everyone has their back to each other, unable to see the materials, but still able to hear the mission director.

Materials Needed

- Manila folders, 1 for each team member
- Toy kits, 1 for each participant (Use LEGOs, K'NEX, or other toys you may have at home, making sure that each kit has the same number and type of pieces. The complexity is up to you!) If you need inspiration for what structures you will create as mission director, look here:

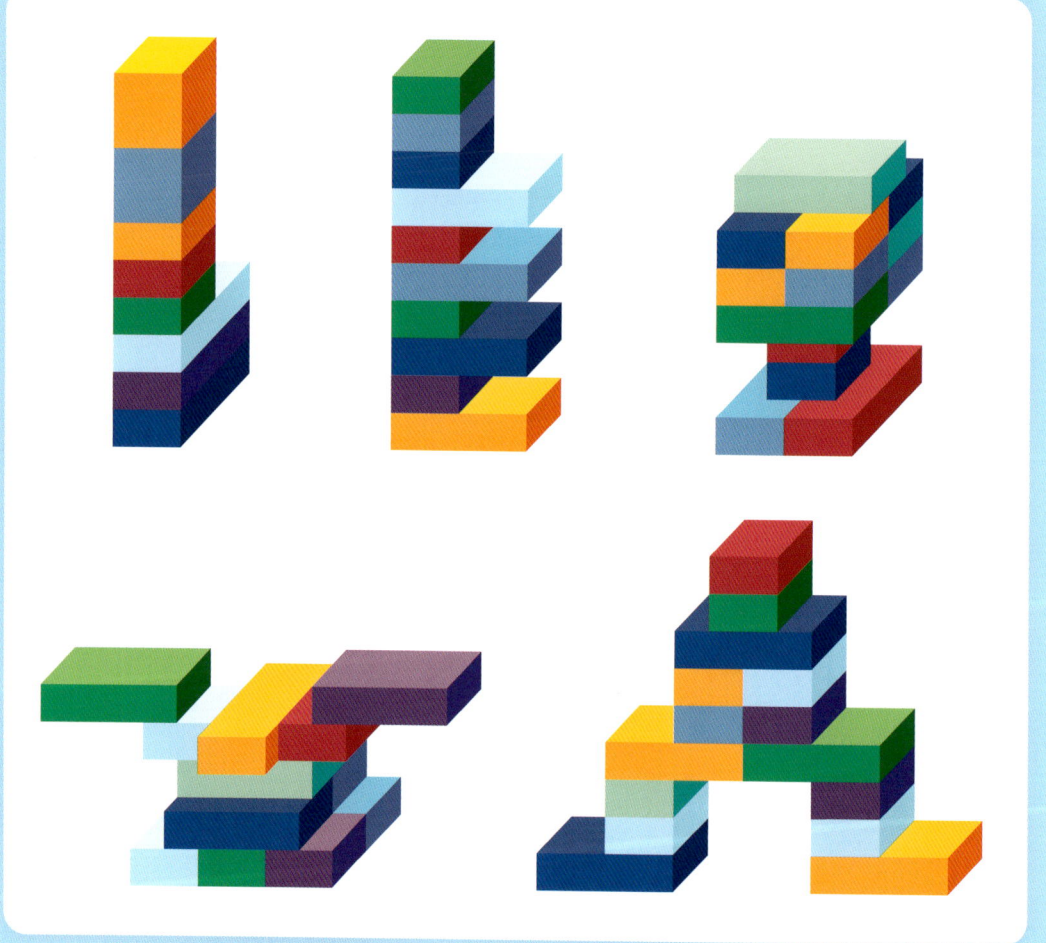

- If you do not have such toys at home, that's okay. Like astronauts, you can improvise:

- Locate paper and scissors. Ask an adult for help.

  1. Cut out a set of ten shapes, one set for every team member. You can play with the color and size of the shapes but, remember, every set must be the same.

  2. You can find examples below. Print them if you have access to a printer, or simply cut them out if you don't.

  3. Assemble the kits and distribute them.

The Experiment

1. Set up manila folders to block the view of the project kits and materials by others, or have participants move to where they cannot see each other's work.

2. With the kits distributed, the mission director takes a few minutes to assemble a structure.

3. The mission director now describes the shape, orientation, and even how to build the structure. The amount of detail is up to you, though keep in mind that the goal is for everyone to have the same shape in the end.

4. How you check the astronauts' work is up to you. A simple method is for the mission director to ask everyone if they believe they have the right shape. Then, have everyone close their eyes and lift their manila folders. See how close or far off they are. Replace the folders and continue to describe the structure, if necessary, until everyone has the correct shape.

5. People may get frustrated, but that's when you know you are learning! Time yourself during each session. Rotate who acts as mission director. Create new structures each time.

If you feel like doing more:

Chart Your Data:

Build Session	Time for All Team Members to Complete	Average Accuracy of Team Builds	Poor, Fair, Good, Excellent
Build 1			
Build 2			
Build 3			
Build 4			

Each time you play the game, note how long it took to see if you are making improvements as everyone learns how to better collaborate.

Life Sciences

Zoology	**Biology**	**Neurobiology**	**Genetics**
"How do animals know when it is time to migrate?"	"How do organisms grow and develop?"	"What is a hormone?"	"Why are identical twins identical?"
Botany	**Marine Biology**	**Entomology**	**Anatomy**
"How does photosynthesis work?"	"How do some fish swim in schools?"	"Why do bees buzz?"	"How do bones grow?"

Life Sciences

Natural sciences involve studying the natural world and physical world. When considering living things, this includes such disciplines as biology and zoology, as well as anatomy, botany, entomology, and neurobiology, plus ecology, genetics, marine biology, paleontology, and virology.

Eugenie Clark
1922–2015

Marine Biologist and Ichthyologist

Temple Grandin
1947–Present

Animal Behaviorist

Corina Newsome
1993–Present

Ornithologist and Animal Scientist

Vernard Lewis
1951–Present

Entomologist

Natasha Tilston-Lunel
1984–Present

Virologist

"I don't work at something because I think it's important. I work at things that, to me, are interesting."

—Eugenie Clark

Quote for YOU

·······························

Dr. Clark says, "Not many appreciate the ultimate power and potential usefulness of basic knowledge accumulated by obscure, unseen investigators who, in a lifetime of intensive study, may never see any practical use for their findings but who go on seeking answers to the unknown without thought of financial or practical gain." However, while that is true, these scientists live fulfilling lives investigating their passions and sharing their knowledge with the world!

Eugenie Clark

Who Is Eugenie "Genie" Clark?

Eugenie "Genie" Clark was born on May 4, 1922, in New York City to an American father and a mother of Japanese descent. When she was nine years old, she and her mother visited the New York Aquarium. Later, her mother bought her a fifteen-gallon fish tank. These experiences, plus her Japanese heritage and the important role that the sea played in her culture, set the stage for Clark's interest in the underwater world. Clark's affinity for fish and aquatic wildlife led her to Hunter College, where she earned her bachelor's degree in biology, and later to New York University, where she received her MA and PhD degrees in zoology. After graduating, she was awarded a Fulbright scholarship and headed to the Red Sea to study the wildlife that lived there. After completing that program, she published an international bestselling book based on her experiences and findings.

During her career, Clark learned how to scuba dive, and she had many awesome experiences under the ocean's waves, such as riding on the back of a whale shark. She also made a number of important discoveries including new

species of fish, like the belted sandfish that can change its sex. Additionally, she worked to dispel myths about sharks. Specifically, Dr. Clark's drive to help clear up misunderstandings about sharks—such as that they are brainless, cold-blooded killers—inspired her to study sharks even more. For instance, funding from the Vanderbilt family gave Clark a chance to open and direct the Cape Haze Laboratory (later renamed the Mote Marine Laboratory), where she investigated shark behavior. Through her studies, she learned additional information about sharks including that some species can be trained to push buttons and hit targets for food. Further, her explorations led her and her team to discover "cleaning stations," or caves, deep under the water where dozens of usually ferocious requiem sharks slept while remora fish cleaned them.

Although the work she did made her one of the leading ichthyologists and shark experts, there were many obstacles she had to overcome. Staring down a wild shark in the depths of the ocean might seem scary, but not for Dr. Clark. Instead, being a woman in a male-dominated field, as well as the prejudice she faced during the time of World War II because of her Japanese ancestry, were among the challenges that made it difficult for her to succeed. However, she did not let these things stop her and, instead, she

Questions for YOU

1. Is there something you like to do for many hours of the day whenever you can—like observe fish at an aquarium or watch a shark swim?

2. What are some other things you like to do and what kind of scientific field might allow you to do those things as a career?

Science YOU Can Explore

- **When was the last time you visited an aquarium?**
  - ☐ Research an aquarium near you. Which species of aquatic animals live there?
    - Identify the different types of programs the aquarium offers that might interest you.
- **Go observe!**
  - ☐ Take a notebook, some pencils, a camera, and your curiosity.
    - Observe the different species of aquatic animals that live in the aquarium.
    - Take notes from the information on the placards, posters, recordings, and tour guides.
    - Draw some of the different species you see.
    - Jot down questions you have about the species you observe.
- **If there is an ichthyologist—also known as an expert on fish—nearby, ask your questions. They may be able to answer them or tell you where to find out for yourself!**

continued to pursue her dreams fearlessly.

Dr. Eugenie Clark became a professor of biology and zoology at the University of Maryland in 1967. She went on to write three books, publish over 150 scholarly and popular articles, narrate documentaries and films, write for *National Geographic*, travel the world talking about science and ocean conservation, and spread the word about sharks, what they are capable of, and their importance to the ocean ecosystem. She conducted more than seventy dives using submersibles, received numerous awards, and was given the name "The Shark Lady" for her expansive body of work. Dr. Clark died of lung cancer–related complications on February 25, 2015, after more than six decades working as a groundbreaking marine biologist.

Fish

What do you call a cold-blooded animal that has a backbone, lives underwater, and uses gills to breathe? A fish! These animals are often covered with scales, have fins that they use for swimming, live in bodies of fresh water or in the ocean, and while some fish give birth to live young, most lay eggs. There are more than thirty thousand different types of fish, and these animals are studied by ichthyologists, or fish scientists. Sharks are one type of fish, and they were Dr. Clark's specialty!

Area of Scientific Interest
....

Animal Behaviorism

Advice for YOU
....

Dr. Grandin believes that all different types of minds are needed in all different fields. Those who think in words, those who think in pictures, and so on. So, encouraging different types of thinking is important.

She also says to go out and look for opportunities—don't just wait for them to come to you. Visit science museums, look out at the world, take flower buds apart and see how they grow, get out and see things, watch satellites go by in the sky, look at bugs. Go out and explore! Discover new things and try something new!

Temple Grandin

Who Is Temple Grandin?

Mary Temple Grandin, known as Temple Grandin, was born on August 29, 1947, in Boston, Massachusetts. She grew up surrounded by supportive adults, such as her mother and grandfather. Grandin's mother encouraged her engagement in art and supported her daughter's development. Grandin's grandfather, who was an engineer, patiently answered many of the questions Grandin had and taught her a number of new things. Experiences such as these, as well as science teachers who motivated her to study hard, all got her interested in science—as a matter of fact, one of her favorite books as a kid was one that focused on famous inventors. She also enjoyed drawing pictures

and could do this with ease. After graduating from high school, Grandin went on to complete her bachelor's degree in psychology at Franklin Pierce College, and earned her master's degree from Arizona State University, and her doctorate degree in animal science from the University of Illinois Urbana-Champaign. Using her knowledge and expertise, as well as her charisma and confidence to try new things, she sought out opportunities to write for state farm journals, and even gave her card to an editor, which led to her chance to write for national cattle magazines.

Dr. Grandin then moved on to design facilities that handled cattle. She used her mind and allowed herself to consider

 Question for YOU

What fascinates you? Is it rain gutters? Sticks? Bugs? Figure out what fascinates you and explore it.

how animals view the world, and then used that information to design facilities that helped support the animals living there. Her work continued to make positive changes for the welfare of animals as she created and supported the implementation of animal welfare auditing systems, which led to the application of better treatment of animals within large companies such as McDonald's, Wendy's, and Burger King.

Despite her many accomplishments, Dr. Grandin came to realize that her brain worked a little differently from many others' because she thought in pictures instead of words. For example, she learned to speak later than most children and, eventually, she was diagnosed with autism. But Dr. Grandin shared that an even bigger challenge she faced was being a woman in the cattle industry, which often included only men. This led to many obstacles she had to overcome. However, her determination paid off, and she continued to solve problems related to how an environment or objects within that environment might affect the animals, facility criteria, and modifications for animal management— she even worked to help children feel included within educational environments, regardless of their way of thinking!

The treatment Dr. Grandin experienced pushed her to become really good at what she was able to do. Not only did her inventions make lasting improvements related to the treatment of animals, but she also went on to advocate for

Science YOU Can Explore

....

- Go out and make a kite. Find materials and put them together. Then, attach different types of tails to the kite to see whether they change how the kite flies.
- Get different-shaped bubble wands (circular, square, star, swirly, etc.) and see what happens when you blow bubbles with them.

individuals on the autism spectrum. She speaks all over the world about how all types of thinking are important and that adults must encourage children to follow their interests no matter what, as well as how important it is to develop and implement educational programs that support all types of thinkers! Currently, she is a professor of animal science at Colorado State University.

Did YOU Know About . . . ?
Autism Spectrum Disorder (ASD)

Autism spectrum disorder (ASD) is a disability that happens due to developmental differences within the brain that may have occurred for a variety of reasons—some known, such as genetic conditions, and others unknown. These differences lead individuals with ASD to behave in ways that may be considered different from typical behavior. For example, individuals with ASD may struggle to communicate or interact with others, to learn, or even to speak. Additionally, some people on the spectrum might have only slight differences, whereas others may have very severe differences that may make it hard for them to care for themselves or to connect with others. Dr. Grandin is an example of someone who has ASD, but it has not stopped her from excelling in the field of science.

Area of Scientific Interest

·······································

Entomology

Advice for YOU

·······································

Dr. Lewis says that the world is changing. It's getting closer together with the internet and other tools. So, it's okay to be different! Remember that! But for all of us to live together, we have to stop shouting and screaming. We have one shot on Earth. He also says to keep a notebook with you and draw pictures. This is priceless.

Vernard Lewis

Who Is Vernard Lewis?

Vernard Lewis was born February 22, 1951, in Minneapolis, Minnesota, and he was born to study bugs. He was adopted by a loving family, and at a very early age, moved from Minnesota to Fresno, California. Lewis would catch bugs and put them in jars, and would often spend his time during kindergarten recess chasing grasshoppers and placing them down harvester anthills, among other things. His grandfather, who also had an affinity for being outdoors, encouraged Lewis to explore nature and insects, and his grandmother taught him early on about the importance of discipline and having a strong work ethic.

All through elementary, middle, and high school, he was a good student and he enjoyed learning. This, plus his love of insects, led him to pursue a career in science. After hearing from a counselor that the University of California, Berkeley, (UC Berkeley) was one of the best schools in the country, Lewis set his sights on attending. Although he did not get in at first, he kept working at it and went to a community college to build his knowledge and academic skills. After excelling there, he got a scholarship to Berkeley and completed his bachelor's degree in agricultural science, as well as his master's and doctorate degrees in

Questions for YOU

1. Do you know what an insect is? Is a spider an insect? Is a roly-poly an insect? Go out and explore insects and learn about them while you do it!

2. Do you have a favorite insect?

entomology. He was the first in his family to attend and graduate from college.

Dr. Lewis's work ethic paid off and allowed him to not just study insects, which he loved, but also to become the first Black professor of entomology at UC Berkeley and a founding member of the United Nations Global Termite Expert Group—but his path was not without challenges. When he was in high school, a counselor told him he was not smart enough to go to college. This was the same counselor who told him that UC Berkeley was one of the best schools in the world. Instead of letting the counselor discourage him, he used those words as fuel and for inspiration, which set him on his path to not only graduating with three degrees from UC Berkeley but also becoming a prominent faculty member there. Additionally, due to lower percentages of people of color in STEM fields (science, technology, engineering, and mathematics), Dr.

Science YOU Can Explore

- ● **Conduct a bug race.**
  - ☐ Carefully collect some bugs, line them up at the same spot, and race them! See which bug gets to the finish line first! Then gently put the bugs back where you found them.
- ● **Collect your observations.**
  - ☐ Go outside and look around. No two things look alike (trees, rocks, etc.).
    - ▪ Do you see any birds? Any bugs?
    - ▪ Write down what you see in your notebook. Sketch what you see.
    - ▪ Date your page and put the time, too.
    - ▪ Come out the next day and observe the same area. Are there differences? Similarities?
    - ▪ Try the same spot early in the morning or late at night—with your parents' or other adult's permission.
    - ▪ Put a black light out—what happens?
    - ▪ Look at insects through a magnifying glass. What do you see?

Lewis rarely encountered teachers or researchers who looked like him. But he did not let that stop him. Instead, he became a role model for so many who have an interest in pursuing careers in science!

Through his work, he traveled the world learning and teaching about insects, helping solve problems related to insects and infestations, and finding ways to use other means to control pest problems that did not require dangerous chemicals. Further, he continued to solve problems related to what was considered a good or bad bug, how to know the difference, and what to do about them. Not only this, but he passed on his passion for insects to his students; even children were encouraged by his love of bugs, including people who met him as a young child, and who went on to become PhDs in entomology themselves! Although Dr. Lewis has since retired from his long and fruitful career, he is considered one of the world's leading experts on termites and an all-around nice "Bug Guy."

Did YOU Know About . . . ?
Termites

Termites are insects that are often pale in color with a soft body. They are social bugs that live in large groups that include queens, workers, and soldiers. Although they may look like white ants and have similar social structures, they are actually more closely related to cockroaches. Also, termites eat wood that they find, which can include dead or rotting trees and plants, houses, and furniture.

Advice for YOU

Newsome says that depending on your background, you may feel like everything is brand-new, but that is OKAY; you can do it. There is always something new to notice! Look it up! Learn about it! This can give you an idea of what's around you.

Corina Newsome

Who Is Corina Newsome?

Corina Newsome was born on April 3, 1993, in Philadelphia, Pennsylvania. She knew as early as four years of age that she wanted to be a scientist who studied animals and bugs. Her passion came from her father, who passed away when she was young, and her mother, who enjoyed being outdoors. As a child, Newsome found herself reading wildlife magazines, watching nature documentaries, and participating in activities focused on animals. When she was twelve, she began volunteering at a veterinarian's office and continued to do so until she was eighteen and graduated high school. Although she knew she had a major interest in animals, she realized that she did not want to be a veterinarian. She also did not know of many other careers that allowed people to work with animals—until she met a Black zookeeper named Michelle and got an internship at the zoo. Thanks to Michelle, not only did Newsome come to see other Black people working in fields related to animals, but she was also exposed to a wider range of careers that would allow her to follow her passions. Newsome then began her academic career and pursued her bachelor's

Questions for YOU

1. What have you noticed in the spaces you inhabit? Your front yard, backyard, schoolyard, etc.?

2. What wildlife have you seen? Could there be any that, maybe, you haven't noticed? Look closely at plants and bugs all around you—and take notes.

degree in zoo and wildlife biology at Malone University.

After graduating, she worked as a zookeeper for four years, and although she interacted with many different animals, birds were her favorite. She loved training birds, doing public education utilizing different species of

Science YOU Can Explore

- How do the birds you see change depending on the environment you are in? Think about different environments, such as roads, beaches, and forests.
- What birds do you see when there are lots of roads/buildings versus forests, trails (away from roads), and places where there is less human activity? Understanding how the species of birds change depending on where they are, and knowing how the world you live in affects the world we share with animals is very important.
- You can apply this to anything in nature, and how different things live and grow in different environments. Think about trees, grass, roads, buildings—any variable you can think of! And think about how the environment affects the things living within it and around it.
- Have you ever reached out to a scientist?
  - ☐ Go out and connect with scientists. Kids can reach out early to begin learning about different fields, especially because scientists often love talking to kids and their parents about science.
    - If you have a particular interest, look up scientists you'd like to connect with. Have a parent or teacher reach out—they really love talking about what they do to encourage the next generation of young scientists.
    - This is helpful because it lets you know so much more about the field and the many options there are for you.
    - Maybe you don't know exactly which area, but you know your interests. Talking to scientists can help you know when you don't like something as well as when you do. You can learn that very early on and explore other options.

birds, as well as learning how to protect them and their habitats. Newsome then left zookeeping and returned to college. She attended Georgia Southern University and completed her master's degree in biology with a focus on ornithology—the study of birds. She visited various environments and ecosystems that were so different from what she had experienced while living in the city, and she learned a tremendous amount and had amazing encounters. Newsome's specific area of study included observing salt marshes and studying the seaside sparrow. During her studies, she actually observed a behavior that had never been seen before—a fish leaping out of the water into a nest and eating birds—and published her findings!

Although she had finally found her calling, pursuing her dreams meant overcoming many obstacles. Yes, she knew early on that she had an interest in animals, but she had no idea until much later about the types of careers she could explore related to this passion. She also did not see very many Black scientists working in the fields of animal science, which made it hard for her to picture herself working as a scientist. She knew math would be an important part of her career, but she understood that it was a challenging subject for her. Finally, being from a city sometimes made going to brand-new and unfamiliar spaces, like marshes, daunting to her. However, she did not let any of this keep her from reaching her goals! She worked hard to improve her math skills, pushed herself outside her comfort zone to have new experiences, and now she is an example to young children everywhere that you can be a Black, female ornithologist and be awesome!

Newsome now works for the National Wildlife Federation and is dedicated to studying and solving problems related to how the environment, and changes within it, affects birds and bird predators. Her goal is to help preserve natural environments and to spread the word about ways to help vulnerable people and wildlife stay safe from environmental threats.

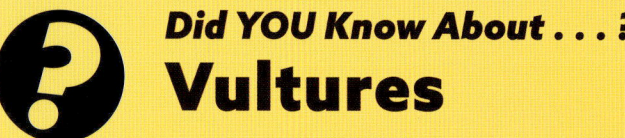

Did YOU Know About . . . ?
Vultures

Corina Newsome says vultures are a really underappreciated group of birds. If these birds were not doing what they do, which is eating dead animal carcasses, we would be in trouble. Vultures have really cool bodies that allow them to eat dead and rotting material. Their stomach acid is up to one hundred times stronger than ours, which allows them to neutralize and kill any dangerous stuff in dead animal carcasses. It is important that

carcasses be disposed of because the bacteria and other nasty things in dead carcasses are deadly to us if we ingest them, if they get into the soil, and so on.

Newsome says the next time you see a large, dark bird soaring above the skyline, there is a good chance it's a vulture. Tell that vulture hi and thanks!

Area of
Scientific Interest
....
Virology

Quote for YOU
....
Dr. Tilston-Lunel says, "Hold on to
your passion as tight as you can.
You are going to have so many
instances where that's crushed
and challenged. Just continue
believing in yourself and never
take rejection personally."

Natasha
Tilston-Lunel

Who Is Natasha Tilston-Lunel?

Natasha Tilston was born on October 19, 1984, in London, United Kingdom, to an English father and an Indian mother. When she was very young, her family moved to Bengaluru (Bangalore), India, where she lived for fifteen years. During her childhood in India, she noticed many individuals who were not well-off or who had medical issues, and she became interested in becoming a medical doctor because she wanted to help. She would also watch countless documentaries with her grandmother about nature and doctors traveling to different places to help people who lived in impoverished situations. Then, in high school, she was fortunate enough to have a wonderful biology teacher who made science fascinating to her. These experiences, plus her ingrained desire to help others around her, set her on the path to study science. She completed her undergraduate work in microbiology, zoology, and chemistry and graduated from Bengaluru (Bangalore) University in India in 2006. Then, she moved back

to the United Kingdom with plans to apply for medical school, but had to wait a year and so decided to pursue a master's degree in environmental science instead. This led her to meet a geospatial scientist and together they mapped how a mosquito-transmitted disease called chikungunya virus could spread using climate data. She was hooked and wanted to learn more!

After completing her master's degree, she went to work first at the London School of Hygiene and Tropical Medicine, where she used mathematical modeling to understand how influenza virus spreads, and then she went on to work in a laboratory at Queen Mary University of London, where she learned about Mycobacterium tuberculosis. In 2011 she got accepted into a doctoral program for virology at the University of St. Andrews, Scotland. This program gave her a chance to engage in a laboratory setting studying viruses. Once she completed her degree she moved to Boston to start her postdoctoral training. Her journey took her to the Center for Vaccine Research at the University of Pittsburgh, where she continued studying viruses such as the measles virus and canine distemper virus. However, during 2020 when the COVID-19 pandemic hit, she and her colleagues started working on a vaccine for severe acute respiratory syndrome coronavirus 2 (SARS-COVID-2) SARS-CoV-2 is the deadly virus that was killing millions of people worldwide.

Questions for YOU
....

1. Do you think there are any viruses in the ocean? If so, how many viruses?

2. How big do you think a virus is?

3. How many viruses currently exist on Earth?

4. Can you name some viruses that can make you sick?

Science YOU Can Explore

- **Have you ever been sick? What did it feel like? Do you remember the time it took from falling ill to feeling better? Jot down some ways you felt when you were sick.**
- **Now, from a safe distance, observe someone who is sick.**
  - ☐ What are their symptoms?
  - ☐ What does their skin look like?
  - ☐ Their eyes?
  - ☐ What is their body doing?
  - ☐ Do their symptoms seem like anything you've experienced before?
- **After observing, look up some of the symptoms and learn a little bit more about what is behind some common illnesses. Here are some questions to research:**
  - ☐ What causes:
    - The common cold?
    - The flu?
    - COVID?
    - Pink eye?
    - Chicken pox?
  - ☐ What are some symptoms for each of these illnesses?
  - ☐ What are the treatments used to fight the illness?
  - ☐ What can people do to avoid catching these sicknesses?

In addition to working as a virologist, Dr. Tilston also had other challenges she had to overcome, such as being a woman in the field of science. For much of her early career, people told her medicine and science were too difficult for women, and they tried to make her change her mind. When she started applying for doctorate programs, she received many rejections that kept her out of the programs. However, she did not let any of this stop her, and now she no longer allows the opinions of others to keep her from reaching her own dreams. She has received major funding from the National Institutes of Health (NIH) to conduct research, and has started her own lab at the Indiana University School of Medicine, where she studies viruses known as bunyaviruses. These viruses can make both humans and animals very sick. Her work studying how viruses evolve is important to help the medical field understand the consequences these viruses can have on humans and animals, and to help develop vaccines and therapies to prevent the spread of these viruses and treat the diseases they cause.

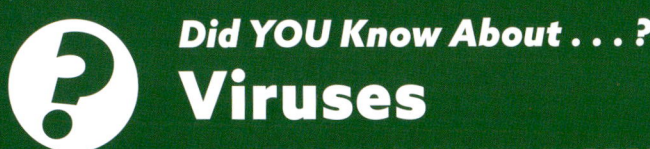

Did YOU Know About . . . ?
Viruses

The word "virus" comes from the Latin word for "poison" or "venom." Specifically, a virus is a rod- or sphere-shaped infectious agent that can be made up of DNA or RNA and is coated by a protein. These agents are smaller than bacteria. Their tiny size allows them to infect animals, plants, and bacteria, multiply, and cause a host of diseases such as the common cold, influenza, and even mumps, measles, and chicken pox in humans. Scientists have developed vaccinations as ways to protect against viruses. Virologists are scientists who study viruses.

"The only place on earth where immortality is provided is in libraries. This is the collective memory of humanity."

—Temple Grandin

Science Activities

Life Sciences

Activity 1:
Virus!

Activity 2:
How We Grow

Activity 3:
Soundscape Ecology

Activity 1
Virus!

Virologists study and specialize in viruses. They diagnose and treat viral infections or develop vaccines or antiviral medications. Understanding how our bodies fight infection can determine how best to respond to a viral outbreak.

Viruses have a protein coat called a capsid. Inside the capsid can be two types of genetic material—either RNA or DNA. A virus multiplies by entering human cells, taking off its coat, and inserting its DNA into the cell's DNA. This causes the cell to make many copies of the virus until the cell explodes, spewing more viruses to infect other cells.

But our bodies have antibodies that help stick to the protein ends (antigens) and mark the germs for destruction. Your body can produce millions of different antibodies in small numbers. And your immune response will be faster and stronger if you have had previous exposure to the germ or had a vaccine to prevent against it.

Antiviral medicines do not kill viruses but, instead, inhibit viral growth. By slowing the replication of a virus, your body builds the antibodies needed to fight that virus.

Materials Needed

- 21 plastic cups
- Stopwatch
- Pair of oven mitts
- A pair of socks
- Pencil
- Notebook

The Experiment

Your task is to build a tower as quickly as possible using all the plastic cups. This represents the speed and ease of a virus multiplying.

Round 1: Tower Building

1. Using a stopwatch, build a tower with all twenty-one plastic cups.

2. Record your time.

Round 2: Tower Building Variations

1. Using a stopwatch, build the same tower with all the plastic cups while wearing a set of oven mitts. They represent the antiviral medication.

2. Record your time.

3. Repeat the steps above with socks on your hands, representing a different variation of the antiviral medication.

4. Record your time.

Record Your Results in a Chart

Trial #	Number of Cups Stacked in 30 Seconds	Time Completed
Trial #1: without oven mitts		
Trial #2: with oven mitts		
Trail #3: with socks on your hands		

Which trial was easiest to build the tower of cups successfully?

Why would it be important to have diverse ways to treat viruses?

How We Grow

All plants and animals grow. Insects have two main ways of growing, called *complete metamorphosis* and *incomplete metamorphosis*. With complete metamorphosis, the young insect does not look like the adult. In incomplete metamorphosis, the young insect looks like a miniature version of the adult.

Insects aren't the only animals that go through complete metamorphosis. Frogs do, too. Frogs start out as eggs that hatch into tadpoles. Over time, the tadpole grows and changes to look like an adult frog. Like frogs, young insects have different names from adults.

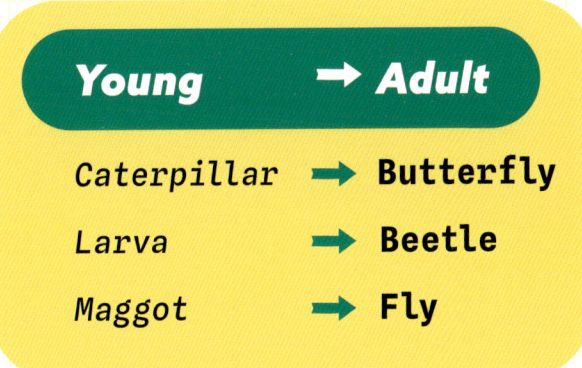

Young	→ Adult
Caterpillar	→ **Butterfly**
Larva	→ **Beetle**
Maggot	→ **Fly**

The Experiment

1. Brainstorm alone or with a partner about any animals you might already know that go through complete or incomplete metamorphosis.

2. If you can, observe insects at different life stages or, if you can't observe insects, search in a book or online for images of insect life cycles.

3. Make a chart like the one on the next page. In the first column, write the name of the adult insect.

4. In the second column, list whether the insect goes through complete or incomplete metamorphosis as it grows.

5. In the last column, describe the stages of life by comparing the similarities and differences between the young and adult.

TIP: Try to be as descriptive as possible. Remember that detailed observations are important for scientists.

6. Once you are done, try to infer what the pros and cons might be of the two ways insects grow.

TIP: There isn't a right or wrong answer when you are making predictions based on your observations. Scientists will use their predictions and observations to design experiments and determine whether what they thought was correct or needs to be changed based on further observations.

Adult Name	Complete or Incomplete Metamorphosis?	Juvenile vs. Adult Life Stages
Example: Frog	Complete Metamorphosis	The young form called a tadpole has no legs and a round body. The body has a long flat tail that helps it swim. The top is darker colored than the bottom to help it hide. Adult frogs have large bodies and four legs. These large legs help the frog jump and, sometimes, they have webbed feet to help them swim. Adult frogs do not have tails.

Activity 3
Soundscape Ecology

When biologists study a habitat, sometimes they can't always see the animals they want to study—but they can hear them. This is especially true for animals like birds. Recording and analyzing the sounds in a location can give the scientist important information they might not be able to gather with visual information alone. The field of science called "soundscape ecology" studies the relationship of living organisms, humans, and the environment through sounds. It can include animal sounds like frogs croaking, human sounds like traffic, and geophony, or the sounds of Earth, like ocean waves. Studying the sounds of an environment can answer many important questions. Also, listening to an environment at different times of the day or different times of the year can change the soundscape. Think about the dawn chorus, the time when birds sing right before sunrise, or how traffic sounds might increase during local rush hour.

In this activity, you will be making your own soundscape of a local area.

Materials Needed

- Notebook
- Writing utensil
- Watch

- An item to record sound
  Tip: If you don't have a specific audio recorder, or a sound recording app on a phone or other technology, use the camera to record video—it'll capture the sound, too.

The Experiment

1. Decide where you want to record your sounds. Do you want to do it in different locations at the same time of the day, or do you want to record in the same location at different times of the day? If you need to, be sure to ask adults about traveling to a different spot.

2. When you first arrive at your location, write general observations in your notebook. Where is the location you will be recording? What is the weather like? What is the sound level (loud, quiet, or in between)? Do you notice a specific sound that is the most common? What is the date and time? Is there anything else you notice?

3. Sit quietly and record for 5 to 10 minutes.

4. Take your recording home, find a quiet spot, and listen to it carefully. If you have headphones or earbuds for the device, they will help, but they aren't necessary.

5. Make notes about what you hear. What sounds do you notice? How often do you hear them? What do you think is making a specific sound? Did you hear anything when listening this time that you missed when you were out in the environment recording?

If you feel like doing more:

1. Find a soundscape that has been recorded by a scientist by searching on YouTube, Spotify, or bird-specific research organizations, such as the Cornell Lab of Ornithology. Listen to the soundscape you find and write observations about the sounds you hear. Compare this to your own recordings.

2. Pick a habitat, like the rain forest, mountains, beach, etc. Write what you would expect the soundscape for this ecosystem to include.

Social Sciences

Psychology

"How do babies learn to talk?"

Anthropology

"What did ancient humans eat?"

Law & Politics

"Who makes the laws?"

Sociology

"Why do some things cost more than others?"

Archaeology

"How do archaeologists know where to dig?"

History

"Who is the first ruler to ever exist?"

Social Sciences

Social science is an area of science that investigates humans—how they behave individually, in groups, and across time and space. Specific disciplines include psychology and sociology as well as anthropology, history, and politics, and also economics, archaeology, and law.

Michael Blakey
1953–Present

Biological Archaeologist and Anthropologist

W. E. B. Du Bois
1868–1963

Sociologist, Scholar, and Activist

Alison Gopnik
1955–Present

Developmental Cognitive Psychologist

Jane Goodall
1934–Present

Primatologist and Anthropologist

Omar Lizardo
1974–Present

Cultural Sociologist

"The worker must work for the glory of his handiwork, not simply for pay; the thinker must think for truth, not for fame."

—W. E. B. Du Bois

Areas of Scientific Interest

Biological Archaeology and Anthropology

Advice for YOU

Dr. Blakey says to believe in yourself! You are capable of more than you can even imagine. No one else really knows what you are capable of. Also, Dr. Blakey believes it is important to ask questions, go out and take brilliant chances, try new things, and explore your interests. Know that your interests can lead to a long, happy career!

Michael Blakey

Who Is Michael Blakey?

Michael Blakey was born on February 23, 1953, in Washington, DC. His father was a dentist and a leading researcher on endosseous implants, and his mother was an artist with a degree in art and a love of being outdoors. Both his parents valued and had passion for science, and both were vocal, active pillars in the Black community during a time when racial tensions were high. Blakey had always had an interest in science. He and his two brothers participated in the Amidon Plan, an educational program that supported high-achieving students within the southwestern Washington, DC, area. At the age of ten, Blakey knew he wanted to attend Oxford University, get a PhD, and become an

archaeologist. Many experiences he had continued to support his passion for science, such as his excursions with his uncle Kermit, where they dug for and collected Indigenous North American artifacts, to the annual science fair he always participated in—and in which he won a grand prize during his last year of junior high school. Additionally, the opportunity to work closely with a prominent paleopathologist named Dr. Donald Ortner continued to solidify his

interest in science. At the age of fifteen, and under the supervision of Dr. Ortner, Blakey was conducting his own research projects related to diet, dental disease, and the muscle attachments of bone.

Throughout his high school and early college experiences, Blakey explored many other interests, such as music. However, Blakey found his way back to anthropology, strengthened by the experiences he had endured and with a clearer understanding of social problems and the inequities tied to them, such as class, gender, race, and racism. Blakey received his bachelor's degree in anthropology from Howard University and went on to receive his master's and doctorate degrees in anthropology from the University of Massachusetts, Amherst. Further, Dr. Blakey continued conducting research related to the bones of past humans, using what he found as evidence for how those

individuals lived. He traveled to places such as Belize and London to conduct his work. Back in the United States, he led groundbreaking research on the African Burial Ground in New York City, where he worked with a large group of scientists and community members to research, understand, and preserve the dignity of the individuals buried at the site.

But none of this was without challenges. Throughout Dr. Blakey's life and career, he has had to speak up and fight for communities that are often misrepresented and discriminated against, such as Black communities.

Questions for YOU

1. How many bones are there in a human body?
2. What are their names?
3. How can you tell if bones are healthy?
4. Are there any animals that do not have bones?

His work is sometimes argued against or deemed wrong because he uncovered what others may consider inconvenient conclusions. Yet, Dr. Blakey believes that sometimes, difficult truths may lead to enemies, but that should not stop you from sharing your findings, especially because difficult conclusions can have a huge impact on the lives—past, present, and future—of so many. Dr. Blakey's work is an example of this.

Dr. Blakey currently teaches at

Science YOU Can Explore

- **Have you ever explored a bone? Find a magnifying glass or microscope and explore some bones. Try looking at the bones of chicken or fish:**
  - ☐ What do you notice?
    - ▪ How are the bones the same? How are they different?
    - ▪ What is inside the bone?
- **Draw the bones in your notebook and write about your conclusions.**
- **Do an internet search about bones:**
  - ☐ What are bones made of?
  - ☐ How many bones are there in a dog? A bird? Etc.?
  - ☐ What is the purpose of bones and why are they important?
  - ☐ How do you mend a broken bone?
  - ☐ What can you do to keep your bones healthy?

the College of William & Mary and is collaborating with individuals and solving problems within the science community to craft ethical guidelines for how to work respectfully with human remains and the descended communities that are connected to them. His efforts are ultimately helping many people regain the dignity and reverence they deserve in death, as well as peace of mind for their living relatives.

? *Did YOU Know About . . . ?*
Paleontology and Paleopathology

Paleontology is the study of fossils, or the remains of things that died. Scientists who study these remains are called paleontologists, and they often work to dig up the fossils to study them. Scientists who examine the bones and the remains of ancient humans and animals for data related to health, nutrition, or pathological conditions are called paleopathologists.

Areas of Scientific Interest

Sociology and Activism

Quote for YOU

Du Bois said, "I believe in Liberty for all men: the space to stretch their arms and their souls. The right to breathe and the right to vote, the freedom to choose their friends, enjoy the sunshine, and ride the railroads, uncursed by color: thinking, dreaming, working as they will in a kingdom of beauty and love." He wished for a world that was free of inequities. He also believed children, especially Black children during a time when Black people were treated unfairly, "have a right to know, to think, to aspire."

W. E. B.
Du Bois

Who Is W. E. B. Du Bois?

William Edward Burghardt Du Bois was born on February 23, 1868, in Great Barrington, Massachusetts. Although his family was poor, Du Bois was a happy child who lived with his mother after his father left. He excelled in school and picked up part-time jobs like mowing grass or delivering papers to bring money home to his family. During Du Bois's early life, racial tensions were high within different parts of the United States, and Black people did not receive the same opportunities as white people for education, jobs, homes, or even basic rights. Segregation was widespread but, luckily for Du Bois, Great Barrington allowed Black people to attend the same school as white people. Thus Du Bois attended and excelled at Great Barrington High School, and he was the first African American to graduate from the school. Then, with the support of different members within his community, funds were raised to send Du Bois to college to continue his education. Though Du Bois had his heart set on Harvard, he went to Fisk University, an all-Black college in the southern state of Tennessee.

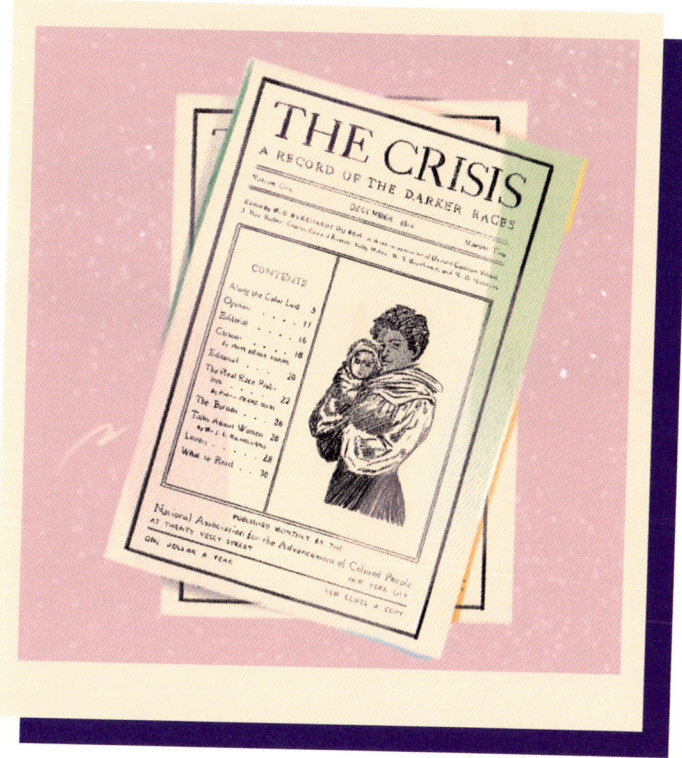

While attending school in Tennessee, Du Bois became much more aware of the difficult circumstances that Black people had to endure in the southern states—harsh realities ranging from not being able to eat in the same restaurants or use the same public buildings or transportation as white people, to white people burning down Black-owned homes and businesses, or brutal beatings and murders of Black people without any consequences. When at Fisk, Du Bois got a deeper understanding of the struggles of the Black community and took actions, such as teaching Black children as a way to help. He graduated from Fisk with honors in 1888, and went on to Harvard where he not only graduated with a second degree, but was also the first Black person to receive a PhD from Harvard. He even went abroad and completed a fellowship in Germany, where he furthered his comprehension of sociology. Du Bois conducted sociological research that allowed him to observe and write about the circumstances and abilities of Black people. He used his training in sociology, as well as his skills, writing abilities, and intelligence to write articles, essays, books, and other types of literature that shined a light on the struggles and achievements of Black people.

Du Bois experienced similar difficulties during his lifetime to those of many other Black people. For example, even after finishing his degree, he had a tough time being hired at white universities. Fortunately, he

Science YOU Can Explore

- Observe the community around you and the people who live there. Are there people who seem to be struggling more than others?
  - ☐ In what ways are they struggling?
  - ☐ What can you do to help?
  - ☐ How can you use science to support your efforts to help?

got a job as a professor and researcher at Georgia's Atlanta University, which allowed him to continue studying and writing about the Black experience. He strived to use his scientific background to educate others about what Black people were capable of and advocate for change. With the publication of his book *The Souls of Black Folk*, he gained popularity as a voice for the Black community, where he encouraged equal treatment and opportunities for Black people. Additionally, he worked with a group comprised of white and Black people to start the National Association for the Advancement of Colored People (NAACP), an organization that worked to gain equal rights for Black people. Du Bois created and edited the NAACP's publication called *The Crisis*, which served as a mouthpiece for the organization and was a way to continue to educate the public. *The Crisis* featured Black art, literature, and poetry, among other things. Du Bois served as the publication's director as well as the director of research for the NAACP, and later a member of its board of directors. His work with the NAACP helped make major strides for equal rights and helped set the foundation for the fight for civil rights.

W. E. B. Du Bois was a staunch advocate for equal rights and treatment for Black people, and he received both acclaim and ire because of it. His work expanded to fighting for Black people and their rights not only in America, but also worldwide, and he even was awarded the Spingarn

Questions for YOU

1. Is there something you think is important to fight for or stand up for?

2. What is it?

3. What ways can you fight for change related to that thing you are passionate about?

Medal for his work. He used his scientific knowledge and background to expand the understanding of the treatment of Black people and to fight for change, and he published many more books and traveled around giving speeches supporting these efforts. Du Bois left the United States and became a citizen of Ghana, Africa, then died on August 27, 1963, after a lifetime of trying to move America toward its widely noted goal: freedom and justice for all.

Did YOU Know About . . . ?
The NAACP

The National Association for the Advancement of Colored People is an organization founded in 1909 by a group of both Black and white people. It was a grassroots organization committed to fighting for equal rights for Black people. Although the organization had a rocky start, it grew and dedicated a tremendous amount of effort to help advance Black people within the United States. For example, individuals who worked with and supported the NAACP helped set the stage for victories in major Supreme Court cases such as **Brown v. Board of Education**, which said that segregation—forcing Black people and white people to be "separate but equal"—was unconstitutional. The NAACP still exists today, continuing the fight for equal rights. You can learn more about the organization at NAACP.org.

Areas of Scientific Interest

Primatology and Anthropology

Quote for YOU

Dr. Goodall says, "I hope some of you will want to learn about animals by watching your pets or the wildlife around your home, or, one day perhaps, wolves or bears, lions or kangaroos. And I hope, too, that you will help us protect them."

Jane Goodall

Who Is Jane Goodall?

Valerie Jane Morris-Goodall, now known as Dr. Jane Goodall, was born on April 3, 1934, in London, England. She was a curious girl who loved animals. As a young child, she enjoyed observing and interacting with animals, and even spent hours barely moving while watching a chicken lay an egg. Her mother encouraged her love of nature, and would often teach her facts about living things as they spent hours together outside in their garden. When Jane was seven, she read the book *The Story of Doctor Dolittle*, about a doctor who traveled to Africa and cared for animals. This book inspired

her lifelong dream to move to Africa and to study animals in their natural habitats. When she was twelve, she even started her own nature group called "The Alligator Club," where she, her little sister Judy, and two friends set up their own events, fundraised to care for old horses, created their own natural treasure museums, and even made their own magazine that included articles, puzzles, and other fun things to read! When Jane

graduated high school, she enrolled in a secretarial school to learn basic skills in typing, notetaking, and bookkeeping so she could earn a living—but she still dreamed of traveling to Africa to live among the animals. That opportunity came when a friend invited her to come live on a farm in Kenya.

While visiting her friend, she learned about a paleontologist and anthropologist named Dr. Louis Leakey who was looking for a secretary. After working for him for a while, he asked her if she'd be interested in going to the Gombe Stream Game Reserve to observe a group of chimpanzees to get a better idea of how they lived. Even though she had no formal training and no college degree, she agreed, and headed to a remote part of Africa to observe the chimpanzees.

While she studied these great apes, she made amazing discoveries that showed how much chimpanzees had in common with humans. For example, she found that chimpanzees engaged in humanlike behaviors including hugging, kissing, holding hands, tickling, taking in orphans, having families, and even engaging in battles and wars over territory. She also learned that many of the chimpanzees showed emotions and had consistent ways of behaving (some were quiet and generous, whereas others were grouchy), which was also similar to human personalities. But one of her most significant findings was that chimpanzees made and used tools! She observed as they would pull leaves off twigs, stick the twigs into termite mounds, and then use the twigs to pull

Questions for YOU

1. How will you make a positive impact on the natural world?

2. How can you help conserve Earth and the living organisms on this planet?

out the termites and eat them. Up to that point, scientists believed that only humans made and used tools, but Jane proved this to be false.

Goodall's findings swept the scientific world by storm. She was offered admittance to Cambridge University in London, where she studied the behavior of animals as she worked on her doctorate degree (even without having an undergraduate degree!). She continued working in the Gombe to study chimpanzees for forty more years, but then switched her focus as she learned that forests were shrinking and animals were being affected as a result.

Dr. Goodall began working to protect nature, to preserve the natural habitats of chimpanzees and gorillas, to spread education about how important it was to protect the Earth and the animals that inhabit it, and to empower youth to make a change in their communities. She opened a nonprofit organization called the Jane Goodall Institute, which led to the opening of additional conservation programs all over the world. Her work has helped protect lands in Africa including the Gombe National Park and other chimpanzee habitats. She also works to help youth understand the important impact they can make, and

Science YOU Can Explore

● Visit the Roots & Shoots website and figure out how you can get involved in conservation efforts: www.rootsandshoots.org.

her Roots & Shoots program gives youth worldwide the tools to make a difference for the better. Jane Goodall continues to travel the world, spreading the word about conservation and showing how every person everywhere can help protect and improve the natural world for all living things!

Did YOU Know About . . . ?
Conservation

The word "conservation" comes from the Latin word that means to "keep safe." In terms of the natural world, conservation means to take care of and protect nature and the environmental resources it provides, including animals and plants, air and water, as well as minerals and the land that covers the Earth. Without conserving natural environments, we are at risk for harming our planet and losing the bounty it provides, as can be seen with animals that go extinct as well as permanent damage to natural habitats. All that will negatively affect future generations.

Advice for YOU

· ·

Dr. Gopnik says that science is not an individual person sitting in their lab! It is a collaborative process that needs many people in the community to think about the same problem and offer their ideas to investigate and to learn about it! Dr. Gopnik also believes that so much of life is left to chance—so go out and have experiences. Get some hands-on experience, after-school experience, science fairs, visit museums—then, go out and design your own experiments.

Alison Gopnik

Who Is Alison Gopnik?

Alison Gopnik was born on June 16, 1955, in Philadelphia, Pennsylvania. As the oldest of six children, she had always been involved in taking care of and engaging with children from as far back as she could remember. Her parents were both college professors, and she spent most of her childhood growing up in Canada. Her fascination for science emerged at an early age when she read a science book on psychology and immediately became interested in the field. Gopnik was a decent student through elementary, middle, and high school, and enrolled in McGill University after she graduated high school at the age of fifteen. Not only did she have an interest in psychology, but she was also intrigued by philosophy, which allowed her to ask big questions about the world, and she wondered why children were often treated as unimportant.

After completing her bachelor's degree in philosophy, she had the opportunity to continue her studies at Oxford University, where she studied under famous psychologist Jerome Bruner. At this time, she ignored her worry of being considered a stereotypical woman pursuing a career in a child-related field, and decided she would pursue a career studying children.

Gopnik had spent countless hours observing very young children and their behaviors, often questioning why and how they did what they did. She enjoyed

Questions for YOU

1. What do you think science is?
2. How do you think science works?
3. How can science be applied to humans and the way we navigate the world?

the challenge that even if you find a lot of evidence that supports your ideas about children's behavior, often, a young child will behave in a different way and prove you wrong! Upon graduating with her PhD in experimental psychology, she went on to serve as a professor and researcher at the University of California, Berkeley, in the Developmental Psychology department, where she currently teaches and conducts experiments investigating children.

However, while Dr. Gopnik's interest in young children led her to academia, she strongly believes academia is not a space designed with working mothers or parents in mind. A major challenge she faced was balancing her life as an academic professor and researcher at a prominent university with being a mother. For instance, during her educational and academic career, she had three children and had to learn to navigate the ups and downs of being a professor while raising children. But

Science YOU Can Explore

- Observe young babies. Are there different sounds they make when they want something compared to other times?
- If you have little brothers and sisters, cousins, or are around young kids on the playground, ask them some questions and get them to explain things. Ask them:
  - ☐ "Why do you think she did that?" after observing the child interacting with someone.
  - ☐ "Do you think different people can like different foods than you do?"
  - ☐ "Why is the moon up in the sky?"
- Listen to what they say in response. You'll see how interesting and unexpected their answers are. Often, their answers are much different from what you'd think, and you can ponder about why.

Dr. Gopnik persisted and did not let those challenges stop her, and thanks to her important contributions to her field, the world has a much better understanding about the minds of very young children— even those who are not able to speak yet! Not just that, she currently advocates for making universities more family-friendly environments for those within them! She continues to solve problems related to understanding how children learn and how this might relate to improvements in computers.

Dr. Gopnik's work has had a profound impact on the field, especially related to concepts such as cause and effect and theory of mind. Her findings have allowed adults to gain a better understanding of how children think and develop, as well as the amazing capacities of babies, toddlers, and very young children everywhere.

Did YOU Know About . . . ?
The Genius of Children

Some studies have shown that if a problem has an unusual solution, children can be better at solving the problem than grown-ups, especially if the answer is not obvious. Further, some studies have shown that kids tend to think science is all about brilliance and genius—you either are that or you are not. But that's not the case! You are doing science all the time in the questions you ask and the investigations you participate in.

Advice for YOU

Dr. Lizardo says that social science makes you curious. Sometimes, you lose the curiosity . . . either because it's so close to you or you have your own ideas. Step back and ask questions—don't take anything for granted—and ask the question WHY? And always have your notepad and pencil and your eyes open so you can observe the things around you!

Dr. Lizardo also believes it's important to think about science in a broad sense and remember that the scientific method can be used to help answer just about any question. Also, remember that science can shed light on things that might be scary or unfamiliar, and it can make those things feel less scary—and that includes how we feel about ourselves and others, too!

Omar Lizardo

Who Is Omar Lizardo?

Omar Lizardo was born on September 7, 1974, in New York City, New York. His parents emigrated from the Dominican Republic, and although he was born in the United States, most of his early education took place in the Dominican Republic. As a child, Lizardo had always been curious, but he never really saw himself as a scientist until much later, when he realized that scientists didn't just study disciplines like physics and chemistry, but they could also study people! Finding his way into the social sciences gave him a chance to explore his curiosities about human nature, and he especially loved how open-ended his field of study is. He returned to the United States as a teenager, and even though he did not know quite what he wanted to do, he knew he had an interest in understanding people. He decided to enroll in Brooklyn College of the City University of New York and to major in psychology. As a senior in his undergraduate studies, he was introduced to sociology and found his passion! After graduating from Brooklyn

College with his bachelor of science degree in psychology, he continued his studies and went on to graduate school in Tucson, Arizona, where he completed his master's and doctorate degrees in sociology.

Dr. Lizardo currently studies a range of topics but especially focuses on the sociology of culture and the study of social networks. In other words, he studies what people like, who they like, and who people are connected to. However, even though he has been successful in his academic career, Dr. Lizardo shared that academia can be hard, especially if you are a first-generation college student, meaning you are the first in your family to go to college. For him, learning to navigate college and academia was like learning a new language. But he did not let that deter him from his dreams. He learned to adapt and kept going, which enabled him to continue exploring and solving problems related to issues of cultural divisions and differences, how people

Questions for YOU

1. What do you wonder about? What is mysterious to you?

2. What motivates you to go out and explore things? Then do it! Learn more about them and gain a better understanding.

Science YOU Can Explore

- **Conduct a network survey:**
  - ☐ Ask six of your current classmates to name three students who were also in their class last year. Write down all of the names.
  - ☐ Draw a diagram showing all of the connections among students in both your current class and the previous classes. And there you have it: a network of classmates across two years!
  - ☐ You can take it a step further and ask them what their favorite movies, foods, and sports are to see if there are any patterns tied to the networks and their favorites.

interact with and are influenced by information they receive, and how to intervene to help people connect with new and more accurate sources of information.

Currently, Dr. Lizardo is a professor at the University of California, Los Angeles, where he teaches sociology and conducts research in his field. Not only has his work shined a light on social networks, how people connect, and how information gets understood and passed on to others, but his work has helped bring together different fields to allow many thinkers to understand and offer perspectives on these topics.

Did YOU Know About . . . ?
Culture

Have you heard the word "culture" and wondered what it means? Culture includes characteristics that describe a group, such as their language, ideas, beliefs, customs, traditions, ceremonies, and rituals as well as the ways in which they interact and behave. Culture varies across the world, and different cultural practices can illuminate the values of each group you come in contact with. For example, some cultures do not look their elders in the eyes because that is a sign of disrespect, whereas other cultures do hold eye contact with elders. Culture is one way to see the beauty in human diversity.

"You cannot get through a single day without having an impact on the world around you. What you do makes a difference, and you have to decide what kind of difference you want to make."

—Jane Goodall

Science Activities

Social Sciences

Activity 1:
Discovering Diets

Activity 2:
Can You Dig It?

Activity 3:
Gone Fishing for . . . Termites?

Activity 1
Discovering Diets

Archaeologists often use bioarchaeology to analyze human biological remains that they find during excavations at archaeological sites. Bioarchaeologists look at bones, hair, and teeth to make hypotheses about the people they are studying.

It is very important for archaeologists to understand the history of the site they are working on. When we think of slavery in America, we most often think of the South. But slavery was not abolished in the northern states until 1804, and some people remained enslaved until the 1850s. Enslaved persons were not allowed to be buried in churchyards in Lower Manhattan, so they created their own burial sites. In the 1980s, bioarchaeologist Michael Blakey analyzed remains dating back to the 1700s within New York City's African Burial Ground. Blakey looked at teeth found in the burial sites to determine how enslaved persons were treated in the colony of New York in the 1700s. Blakey found that many of the individuals buried at the burial ground suffered from malnutrition and extreme overwork.

Materials Needed

- 3 mugs
- 3 cups hot water (not boiling)
- Liquid measuring cup
- Tablespoon
- 3 tablespoons fresh lemon juice
- 3 tablespoons vinegar
- 3 Tic Tacs
- Timer
- Mixing spoon

The Experiment

In this activity, you are going to see how teeth can lose enamel and break down if the person cannot get the right amount of nutrients. Teeth can show how healthy a person was during their life. People who suffer from malnutrition and do not get the vitamins they need often develop *enamel hypoplasia*—the thinning of the enamel on teeth.

- Tic Tacs represent teeth.
- The tooth in the water represents a tooth that has received the right amount of nutrients.
- The tooth in the lemon juice represents a tooth that has not received all the proper nutrients but has received some.
- The tooth in the vinegar represents a tooth that has not received the right amount of nutrients at all.

Please conduct this experiment with an adult present in the room for safety.

1. Hypothesize which tooth (Tic Tac) will dissolve first.

2. Line up the 3 mugs on a table. Carefully pour 1 cup of hot water into each mug using your liquid measuring cup. Ask an adult for help if you need it.

3. Pour the lemon juice into the second mug.

4. Pour the vinegar into the third mug.

5. Drop 1 Tic Tac into each mug.

6. Set your timer for 15 minutes and check back on the Tic Tacs when the 15 minutes are up. Stir the liquid in each cup. Have any of the Tic Tacs dissolved when you returned? (If no Tic Tacs have dissolved, check back in 10 minutes.) Which Tic Tac dissolved first? Why do you think that Tic Tac dissolved first?

7. Look back at your hypothesis. Does it match your findings?

Deduce

Now that you have completed your experiment, let's think of how that relates to bioarchaeology in the African Burial Ground. You found that those teeth that did not receive the proper nutrients (the one in vinegar) dissolved first. That means that those teeth did not have very strong enamel. Many of the teeth remains that Michael Blakey analyzed did not have strong enamel either. More than 70 percent of the individuals Blakey looked at experienced enamel hypoplasia. Remember, enamel hypoplasia means that the enamel has gotten much thinner because of a lack of proper nutrients.

Do you think that many of the individuals Blakey looked at ate a proper diet during their lifetime? Why or why not?

Many of the individuals Blakey studied were not able to get the nutrients they needed. What are some reasons that the individuals might not have had proper nutrients and, therefore, unhealthy teeth?

Tip: Look back at the background history.

Conclusion

Bioarchaeology can show us how individuals lived in the past. It can also inform us of how certain groups of people were not treated with kindness and respect or even basic humanity. It is important to learn from these discoveries and make sure we do not repeat the atrocities of our past.

Activity 2
Can You Dig It?

Archaeologists are like detectives who explore the past by digging up old things like toys, pots, and even houses used by people from a long time ago. They carefully dig and find these artifacts buried in the ground. Then, they study these treasures to learn about how people used to live, what they ate, how they built their homes, and what games they played. It's like solving a big mystery about the past!

In this activity, you will act as an archaeologist, using tools to excavate small toys buried in ice, make observations, and create hypotheses about the individuals who left the items behind. Instead of digging through dirt out in a field, though, we will use ice. Any small toys or objects can be used for this activity. See if you can find items such as doll shoes, toy cars, LEGO objects, and others. And remember to keep small objects away from the youngest explorers (ages three and under).

Materials Needed

- Trays or plastic storage containers
- Small toys or objects (doll shoes, toy cars, LEGO pieces, etc.)
- Freezer space
- Small hammers or mallets
- Toothpicks
- Paintbrushes

- Gloves
- Notebooks
- Pencils
- Magnifying glasses
- Safety glasses
- Someone to explore with— you can hide objects in the ice for them and they can hide objects for you

The Setup

Create the ice blocks for the archaeological dig by freezing them in stages.

Steps for Preparation

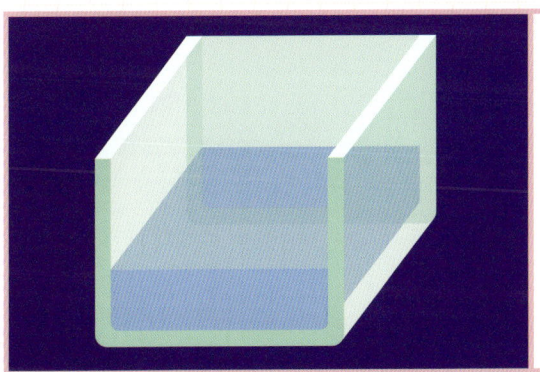

1. Fill a container or tray partially with water and freeze to form the bottom layer of ice.

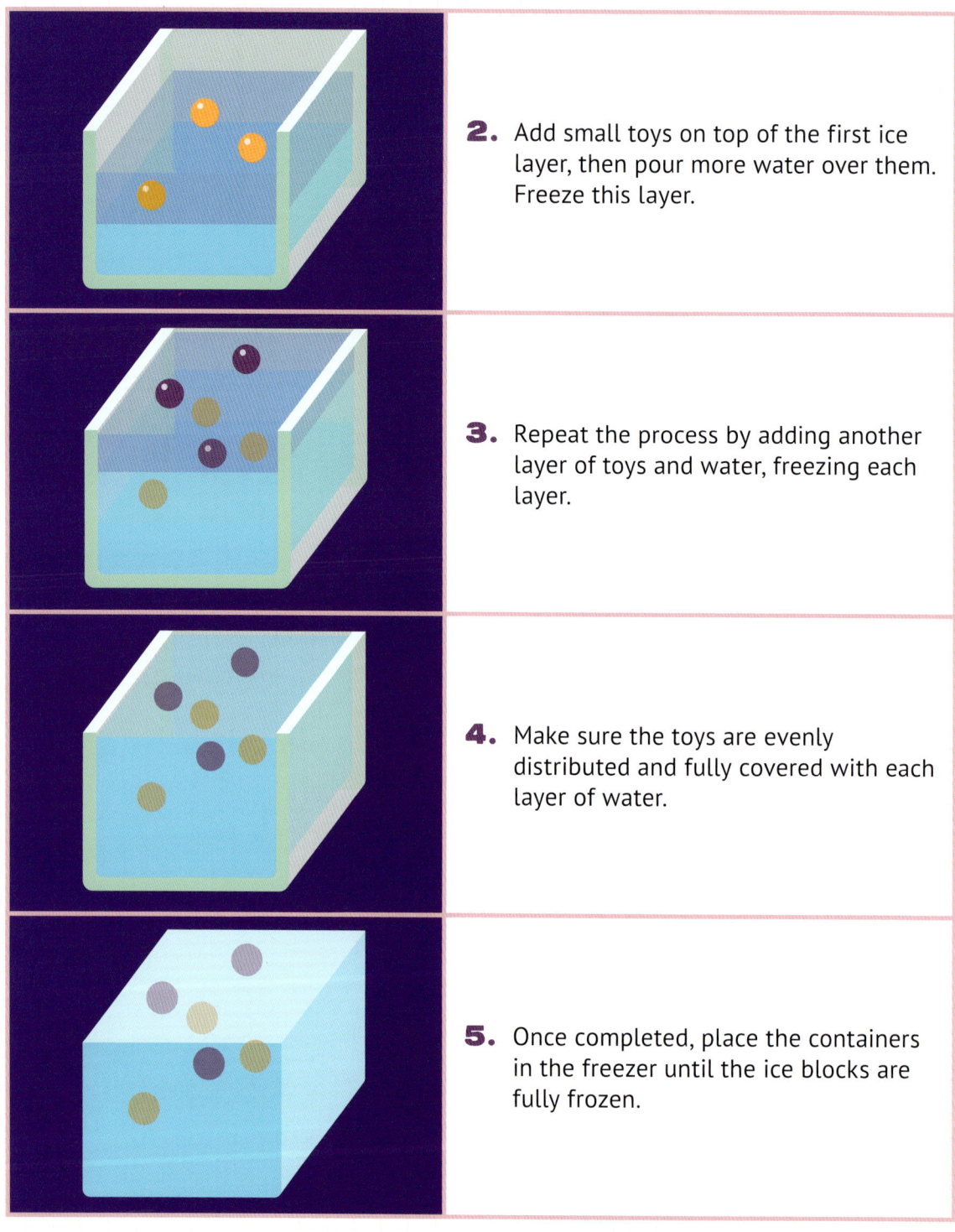

2. Add small toys on top of the first ice layer, then pour more water over them. Freeze this layer.

3. Repeat the process by adding another layer of toys and water, freezing each layer.

4. Make sure the toys are evenly distributed and fully covered with each layer of water.

5. Once completed, place the containers in the freezer until the ice blocks are fully frozen.

The Experiment

1. Use the tools you have to slowly chip away or melt the ice. Uncover the hidden toys buried within. Take your time and observe the changes in the ice as you uncover the toys. Jot down your discoveries in your notebook. Describe the toys, their size, material, and any unique features.

2. Use your imagination to invent an ancient civilization that might have owned these particular items. What could you deduce about the civilization from this dig? What would the people there eat, create, worship, or do for fun? What information about its history or culture might be missing?

3. With these observations, craft hypotheses and creative stories about this imaginary civilization. Share your reasonings, thoughts, and stories with others.

Activity 3

Gone Fishing for . . . Termites?

Jane Goodall's work showed us that animals are smarter and more like us than we thought!

Chimpanzees are our closest living relatives in the animal kingdom. Jane Goodall spent a lot of time with chimpanzees and discovered many amazing things about them, especially how they use tools. Goodall saw chimpanzees encounter termite mounds and use sticks or grass to fish out the termites. The chimps would modify these tools by stripping off leaves or twigs to create a more efficient fishing rod, then stick the tools into the mounds. The termites would bite onto the stick and the chimps could pull them out for a tasty, protein-filled snack. This behavior was important because it demonstrated their ability to plan, modify, and use tools for a specific purpose. For a long time, people assumed that we humans were the only animals with the ability to create and adapt tools for a particular job. However, observing chimpanzees engaging in this behavior—which Jane Goodall did—changed that assumption.

Materials Needed

- Bag of metal paper clips
- Magnet
- A stick or pencil
- Adhesive tape

The Experiment

We can replicate the chimpanzees' tool by using everyday household items. Here's an activity for family participants to create a termite fishing tool using a bag full of paper clips and a magnet.

1. In the wild, termites build tall dirt structures with holes for getting in and out. Fill a paper bag with paper clips. This bag will be our termite mound.

2. Attach a magnet to the end of a stick with tape. Because our paper clips can't "bite" the stick, we'll use magnetism to attract them.

3. Without looking into the bag, slowly put the stick with the magnet into the bag and see how many "termites" you can catch.

While you're creating and testing your termite fishing tool at home, consider these points:

1. Why do you think the ability to use tools is important for animals like chimpanzees?

2. How does the chimpanzees' use of tools compare to human tool use?

3. What tools do we humans use that chimps might be able to use, too?

4. What does the termite fishing behavior teach us about animal intelligence and problem-solving skills?

Medical Sciences

Neuroscience

"What are brain cells made of?"

Ophthalmology

"How do eyes see?"

Pediatrics

"Do vegetables really keep me healthy?"

Internal Medicine

"Why does my heart beat fast when I run?"

Medical Sciences

The medical sciences include many disciplines to help doctors understand, take care of, and cure ailments related to health and disease. The types of medicine are vast but can include such areas as internal medicine, neurology, ophthalmology, and more.

Ben Barres
1954—2017

Neuroscientist

Patricia Bath
1942–2019

Ophthalmologist

Rajani LaRocca
1970–Present

Internal Medicine Doctor and Children's Book Author

Nadine Caron
1970–Present

Surgeon, Researcher, and Professor

Karlyn Beer
1984–Present

Epidemiologist

Area of Scientific Interest

Neuroscience

Quote for YOU

Dr. Barres said, "I lived life on my terms: I wanted to switch genders, and I did. I wanted to be a scientist, and I was. I wanted to study glia, and I did that, too. I stood up for what I believed in and I like to think I made an impact, or at least opened the door for the impact to occur. I have zero regrets . . . I've truly had a great life."

Ben Barres

Who Is Ben Barres?

Ben Barres was born Barbara Barres on September 12, 1954, in West Orange, New Jersey. From a young age, Barres never felt quite comfortable being female, though he was assigned that gender at birth. Despite these feelings, Barres was a stellar student, especially in science and math, two subjects that girls were often discouraged from pursuing.

As someone perceived as female growing up in the educational system, Barres was often denied access to math and science courses, but a summer program at Columbia University in New York offered Barres the opportunity to finally take these classes of interest. Barres finished high school and then completed a bachelor's degree in biology at Massachusetts Institute of Technology (MIT), a medical degree at Dartmouth College, and a residency in neurology at Weill Cornell Medicine.

Throughout his schooling, Barres came to wonder about specific cells within the brain and nervous system— the glial cells—that seemed to be related to many neurological issues. This especially intrigued Barres because, at that point, not much research had been done on glial cells. Dr. Barres decided to leave medicine and pursue a doctorate in neurobiology at Harvard Medical School in Boston, and then went on to a postdoctoral position at University College London to focus on glial cells within the brain. During his studies, Barres discovered a number of astounding findings that contributed important knowledge to this field.

Questions for YOU

1. What is a brain?
2. How does a human brain differ from the brains of other animals and organisms?

Barres went on to become a full-time professor at Stanford University's School of Medicine in California, and started his own laboratory in the Neurobiology department. During this time, Dr. Barres worked with a number of students, encouraging them to ask questions, share their ideas and insights, and push the envelope in their studies.

Dr. Barres faced struggles related to being perceived as female in a male-dominated scientific world. At age forty-three, he came out as a transgender man. He continued to mentor his students and postdocs fiercely, and to speak out about the need for equity in science. His advocacy was an inspiration to many!

In addition to his scientific work, Barres often advocated for equity and inclusion in sciences. He spoke up about why he believed every voice is needed in science, and that diverse perspectives are what help advance every scientific field. He was a staunch advocate, spreading the word about gender inequities within science, the barriers women face, and the need to work to minimize these inequities.

Science YOU Can Explore

- How much do you know about the human brain? Take some time to explore!
- Go to your local library and check out five books about the brain. Look for answers to the following questions:
  - ☐ What is the brain made of?
  - ☐ How big is the average brain of an infant, a five-year-old, a ten-year-old, a twenty-year-old, etc.?
  - ☐ How does the brain communicate with the other parts of the body?
  - ☐ How does the brain learn?
  - ☐ What can you do to help keep your brain healthy?
- Now, ask some of your own questions and explore the amazing human brain!

In 2013, Dr. Barres was elected to the United States' National Academy of Sciences. He was the first transgender member and served as a model for the LGBTQ+ community, illustrating that they do have a place in science and that they belong! Dr. Barres died on December 27, 2017, from pancreatic cancer at the age of sixty-three, but his impact in the neuroscience world, and as an advocate for equity in the sciences, lives on in the people he taught, mentored, inspired, and advocated for, and the generations that will follow.

Did YOU Know About . . . ?
Being Transgender

"Transgender" is a term used to describe individuals whose gender identity, or the way they express their gender, does not match what society considers to be typical for the sex they were assigned at birth. While trans people have existed throughout history and in multiple cultures, education and awareness in recent years have resulted in greater trans visibility across all segments of society.

Quote for YOU

·······························

Dr. Bath said, "The hardest part isn't coming up with the idea—that might be the easiest part, especially if you have a curious, inquisitive, brilliant mind . . . Translating the idea into reality . . . that's the tough part."

She also said, "The most important part of this has been having faith and belief in the power of your ideas. When you're trying to pioneer and advance something that is new, there will be those who say, 'It's impossible,' those who say, 'It can't be done' and one has to have that belief and the personal empowerment to . . . devote the diligence, the study, the work, the sleepless nights, and the lifelong commitment to making it a success."

Patricia Bath

Who Is Patricia Bath?

Patricia Bath was born on November 4, 1942, in Harlem, New York. At a young age, her mother gave her a chemistry set that she loved to play with, and it ignited her interest in science. Bath excelled in school. She also found inspiration in the work of a prominent medical doctor named Albert Schweitzer, who ran a hospital in Gabon in Africa, and helped heal people with deadly diseases. Bath's chemistry set plus the work of Dr. Schweitzer put her on the path to medicine.

When it was time for Bath to begin high school, the high schools in her community did not accept African Americans. Thus, every day, Bath took a subway out of her neighborhood to attend a school in Harlem, but this did not squash her drive. She continued to get high grades in school and won a scholarship through the National Science Foundation to study in a program at Yeshiva University, where she completed a research study focusing on cancer and nutrition—this study even got her recognized by *Mademoiselle* magazine's Merit Award at eighteen years old! She completed community college credits while taking her high school

classes and graduated high school in only two years! After high school, Bath got a scholarship to Hunter College, where she received her bachelor's degree in chemistry. She moved to Washington, DC, to complete her doctorate degree at Howard University College of Medicine, then went back to Harlem to complete her three-year residency at Harlem Hospital Center and a fellowship in ophthalmology at Columbia University.

During her residency and fellowship, she saw many patients and realized that Black people had much higher rates of blindness compared to white people, which she believed was because the Black patients did not have access to quality eye care. These disparities seemed unfair to her—she believed that being able to see was a basic human right. As a result, she created and implemented what she called "community ophthalmology," for which individuals could volunteer their time to learn how to test vision for early detection of serious eye conditions and so patients could receive preventive treatments. After finishing her programs, she moved to Los Angeles to teach ophthalmology at the University of California, Los Angeles, (UCLA) and Charles R. Drew University of Medicine and Science. Additionally, she joined the Department of Ophthalmology at the Jules Stein Eye Institute, where she was the first female ophthalmologist to become a full-time faculty member, and later, served as a head of the residency training program, which she also helped establish. Additionally, she cofounded a nonprofit organization

Science YOU Can Explore

- As Dr. Bath says, "Begin each day by asking a question." Let the answer lead you to another question, and you will discover that learning and knowledge are an infinite playground.

called the American Institute for the Prevention of Blindness. Later, she developed and patented an invention called the Laserphaco Probe, which used cutting-edge technology and science to help remove cataracts from patients' eyes, allowing them to see clearly again and, for some, completely reversing the blindness caused by the foggy eye lens that characterizes cataracts! The Laserphaco Probe was just one of many patents she received for her inventions.

Despite her numerous accomplishments, Dr. Bath was no stranger to obstacles. Throughout her educational and professional career, Dr. Bath faced discrimination related to her gender and race. Often, being a Black woman led to restrictions and underestimation, such as not being able to attend her local high school, nor being able to sit in the front row at some of the educational institutions she attended, as well as colleagues second-guessing her work. But she did not let these challenges

dim her glow; instead, she used them as fuel to "swim faster, swim harder." She found inspiration in the world around her, ranging from mentors to the treatment of patients within the community, and she used this to keep striving for better treatments for eye health. Thanks to her work, thousands of people received treatments for blindness, and some regained their vision after decades of not being able to see.

Dr. Bath taught at UCLA until she retired in 1993. All the while, she continued to educate, write, and spread the word about the importance of eye health. She even became the president of the American Institute for the Prevention

Questions for YOU

1. Always ask "What if?" and then use that to ignite exploration and curiosity.

2. What are some things you can ask "What if?" about?

of Blindness. Dr. Bath died in 2019 at the age of seventy-six after a stellar career impacting the world with her medical prowess, innovation, and desire to improve and preserve the eyesight of people across the globe.

Did YOU Know About . . . ?
Patents

A patent is the right of ownership over an invention someone creates. In other words, it gives an inventor the credit and recognition for the product or process they created. To gain a patent, inventors have to go through an application process and then be granted the patent. Patents can help protect inventions so that others cannot take credit for the work of someone else.

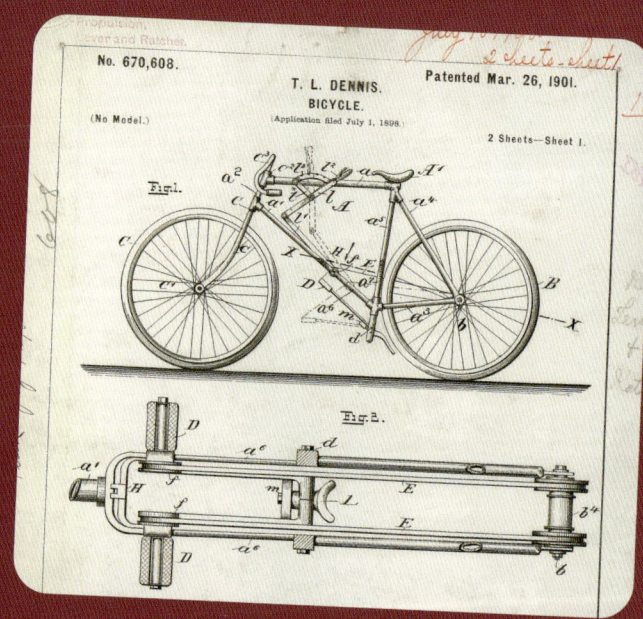

Medical Sciences

Areas of Scientific Interest
....

Surgery, Research, and Teaching

Quote for YOU
....

Dr. Caron says, "Life is like a tree with the roots being the family and your experiences. . . . It's neat for kids to know that they are building a strong trunk right now by being curious, by exploring what they're interested in, by being open to new opportunities, and hearing the positive things said, and letting any negative words roll off them." She also believes that science is a way to practice and learn how to ask questions and find your own answers. If you are open to learning about how to ask questions, and you keep asking questions, you will find the areas that you want to learn about and give yourself the future that you deserve!

Nadine
Caron

Who Is Nadine Caron?

Nadine Caron was born in Kamloops, British Columbia, Canada, on April 23, 1970. Her mother is First Nations Sagamok Anishnawbek and her father emigrated from northern Italy to Canada before she was born. When Caron was young, she loved learning about science, especially at Arthur Stevenson Elementary School, which she attended during her formative years. Her interest in science continued through high school, but it wasn't until she attended college that she realized how curious she was about kinesiology—and that's when she fell in love with the science behind the human body. As a student-athlete on the varsity basketball team, she got to learn about how muscles, tendons, ligaments, and cardiophysiology worked, and then she put her body into action on the court! Caron went to medical school at the University of British Columbia, completed her general surgery residency, and continued on to receive a master's in public health from Harvard University. She also completed a fellowship where she received even more training about how to perform surgeries, especially on cancers within the endocrine system.

Dr. Caron went on to serve as a surgeon, research scientist, and professor. Not only does she use her knowledge and expertise to help improve the health of others, but she also shares this knowledge with students to improve their skills as well. In addition, her ability to conduct medical research allows her to create knowledge and learn things for the first time that can help push the medical field forward. Even with the amazing things she has accomplished across her career, it was not without challenges. Dr. Caron had to endure many disappointments in high school and college, while playing basketball, and even in medical school. However, she does not dwell on these, because she believes that it is the things that make you cry or doubt yourself that actually help you become stronger. These struggles then become a big part of your success!

Dr. Caron has broken many barriers in her field. She was the first Indigenous woman to not only be accepted to the University of British Columbia's medical school, but also to become a surgeon

Questions for YOU

1. What are muscles? How do they work? What about tendons, ligaments, cardiophysiology?

2. What does it mean to be "in shape"?

3. In basketball, why can some people jump and grab the rim while others are better on the ground, playing the defensive role?

4. Why do some people win the battle against cancer and some don't?

5. When it comes to diseases, what causes them, makes them worse, prevents them, or helps improve their symptoms?

Science YOU Can Explore

- There are many things we now know about because of the work of scientists within the field of medicine. For instance, wearing a helmet for biking, skateboarding, or downhill skiing has been shown to decrease the risk of concussion, brain injury, or death. But still, some people do not wear them—even if it's the law. Don't believe this? Do some of your own research!

  - ☐ Conduct a study on helmet-wearing.
    - ▪ Do a survey and write down what you find.
      - ☐ Ask kids in your school:
        - **a.** Who wears a helmet to ride a bike, skateboard, and/or ski?
        - **b.** Whose parents wear a helmet for these activities?
    - ▪ Go out and observe.
      - ☐ Sit in your schoolyard and watch people riding bikes or skateboards nearby. Count how many are wearing a helmet and how many are not.
    - ▪ Take a look at your data.
      - ☐ Did you notice any interesting trends?
    - ▪ Take action.
      - ☐ Ask permission and then post signs in the neighborhood and at school that say "Wear a helmet!"
      - ☐ Two months later, sit in the same place and count the number of people wearing helmets now.
      - ☐ Is there a difference compared to before?
  - ☐ You can do a similar study about crosswalks—who uses them and who does not. Did you notice any interesting trends there, too?

and later, a professor there, too. Not only that, but Dr. Caron was the first Indigenous female surgeon in all of Canada! Despite all her successes, the accomplishment she is most proud of is being a mother. Her daughter brings her so much joy, and to her that is by far the best thing she could ask for. While science and finding answers is of utmost importance, so too is building strong relationships with the people we love—and Dr. Caron has been successful in each of those facets of her life.

Did YOU Know About . . . ?
Surgery

Surgery is its own branch of medicine but also the procedure used for diagnosing and treating a whole range of human diseases and conditions. Specifically, surgery includes using some sort of tool or instrument—such as needles, probes, scalpels, lasers, etc.—to change a part of the human body in some way. The changes can range from cutting, burning, freezing, and suturing to other forms of manipulation to human tissues. Patient safety is of utmost importance when surgery is required, and medical doctors often have to receive additional specialized training and education to become licensed surgeons, or doctors who perform surgeries.

Advice for YOU

·······························

Dr. LaRocca says that picture books are for everybody; for those who don't think picture books should be read by anyone older than eight, *Jeopardy!* champion James Holzhauer credits children's books for his success. He read many children's books as preparation for the show because they are succinct and filled with lots of information! So go out and read picture books to learn new things!

Dr. LaRocca also believes that all the observations you make are so important in science—the best thing you can do is observe and take notes, and maybe come up with an idea of why things are the way they are—a hypothesis!

Rajani LaRocca

Who Is Rajani LaRocca?

Rajani LaRocca was born in Bengaluru (Bangalore), India, on March 19, 1970, but at a young age moved to the United States, where she grew up in Louisville, Kentucky. She enjoyed going to school and reading, and she knew as early as age three that she wanted to be a physician. She was especially influenced by a book called *Ouch! All About Cuts and Other Hurts* and thought about how cool it was that when a body got a small cut or scrape, it could heal itself without even thinking about it! After reading that book, she knew for sure that she wanted to pursue a career in medicine. LaRocca went on to complete her undergraduate degree at Harvard and then her medical degree at Harvard Medical School, where she studied internal medicine. Finally, she completed her residency at Massachusetts General Hospital.

For Dr. LaRocca, and most doctors, getting into medical school is a big accomplishment. Students who wish to apply for medical school must have strong grades through their academic careers, especially in high school and in their undergraduate work, and then must

Science YOU Can Explore

- How do you know whether you're healthy or not?
- Do you notice that you feel particularly good after you do something, like after you run around and play with your friends, get a good night's sleep, or eat when you are hungry?
- Are there certain things that make you feel not so good?
- Things you can do on your own:
  - ☐ Think about your heart! The heart is hidden, but you can hear and feel your heart.
  - ☐ Try this (these are the things doctors did throughout history to figure out whether people were doing well or not so well):
    - Feel your pulse on your wrist or in your neck.
    - When you visit your doctor for a checkup, ask to use their stethoscope to listen to your heart and what it sounds like.
    - Can you see the pulse in your wrist, like when you are running around?
    - Measure your pulse by gently putting two fingers on the inside of your wrist, or by gently pressing the side of your neck under your jaw. What does your pulse feel like at rest? Jump up and down and measure it again.
- Through hundreds of years of observation, doctors figured out the normal range of things, like a heartbeat or how quickly people typically breathe, what their belly should feel like, what their lungs can feel like, and so on. You can rediscover these insights on your own. You can apply these observational skills to all types of things, like your pet! How does your pet react when he or she is nervous, hungry, resting, etc.?

pass a very difficult exam called the Medical College Admission Test (MCAT). Then they must complete several years of challenging coursework in which they learn about human biology and how the body works, until they become experts at the science related to the human body. Thanks to hard work and determination, Dr. LaRocca excelled despite the long hours, numerous tests, and large amounts of difficult information to learn and remember. Throughout her studies, she had to learn medicine-related science, prove she understood it, show she had learned practical skills for helping others, and then do all of these things in real time and real life. After finishing her MD, she had to complete her residency, where she worked as a doctor-in-training for at least three years. After completing

her residency, she continued working at Massachusetts General Hospital, where she practices internal medicine as a doctor for adults.

Even though it is hard work, becoming a doctor is a tremendous feat that is possible for those who persevere, just as Dr. LaRocca did! She loves building relationships with her patients and helping them be their healthiest selves. Further, she is always working to solve problems related to the best ways to treat diseases people get, to keep her patients healthy, learn whether prescribing medication is the best course

Questions for YOU

1. What are you curious about? What are you interested in? What kinds of questions do you have?

2. What do you think will happen when you try something, and how can you go about trying it safely?

of action, and help her patients change their behaviors to improve their health. In addition to this, she is an award-winning children's author and many of her books include scientific and medical-related topics, such as vaccines, DNA, and how the heart works.

Did YOU Know About . . . ?
DNA

DNA, or deoxyribonucleic acid, is a molecule that holds all the instructions for our genetic makeup. DNA is found in the nucleus of the cells of living things, and every type of animal has its own special composition of DNA. But even with these differences, research shows that everyone's DNA is the same size! Elephants are pretty closely related to us, and even their DNA is just about the same size as a human's!

Area of Scientific Interest

Epidemiology

Advice for YOU

Dr. Beer says, "Keep asking questions. Science is about being confident in your curiosity. Know that curiosity will take you to good and interesting places!" We will never have all the answers to all the world's questions, because there are so many unanswered ones. She also says to think about things in your life that make you mad or that seem wrong, unfair, or confusing. What questions can you ask about those things, and how do you think things should be instead? This can lead you down a line of inquiry that is driven by a deeper value that you have (justice, equity, needing things to be right, etc.), and can help you figure out what your passions truly are!

Karlyn Beer

Who Is Karlyn Beer?

Karlyn Beer was born March 23, 1984, in Columbus, Mississippi. Her father was in the military, so she moved to Texas and then Minnesota at a young age, where she lived with her parents and her younger sister. As a child, she enjoyed many activities, including spending time outside and attending school, despite challenges she faced because of a birth defect that led her to have to wear corrective lenses due to poor eyesight. She especially loved it when she got to learn about science and math. When she was in junior high school, she researched a disease called Mad Cow Disease, which sparked her interest in learning more about all types of diseases. It was in junior high that she realized she wanted to become either a Supreme Court justice or an epidemiologist—someone who studies diseases. After trying out the Mock Trial club in high school, she decided epidemiology was the way to go, and thanks to the support of her parents, she began researching, contacting, and interviewing epidemiologists to learn more about what the job involved. This helped increase her interest in the field even more! When Beer graduated from high school, she went off to Cornell University, where she received her BS in microbiology, and then she continued on to the University of Washington, where she received an MS in epidemiology and a PhD in molecular and cellular biology.

During her time in college, Dr. Beer had the opportunity to conduct research and work in laboratory settings studying things ranging from herpetology (the science of reptiles and amphibians), poop, and the brain, to rabies, diet, microbes, and different types of viruses. She studied abroad in South Africa and learned that working with people is much harder than working in laboratories. She even had a chance to take a year off from school and travel, exploring the world and thinking more deeply about the type of science she wanted to focus on as a career. She also realized she really enjoyed working in public health and became what she called a "disease detective," where she had to figure out diseases and their patterns in populations of people. In other words, she investigated who was getting these diseases, when they got sick, where they got sick, and how to stop more people from getting sick. Her knowledge gave her the opportunity to work all across the globe, and to start her own company to bring disease-prevention

Questions for YOU

1. How are you learning about science and scientific information in today's world outside of school?

2. What places have brought you excitement about science?

3. Is there anything you are afraid of related to learning science based on the community you live in?

4. Do you feel there are any barriers to learning and exploring science in your life?

5. What is it like to be a kid learning about science in the age of social media and artificial intelligence?

Science YOU Can Explore

Kids are in school and get sick all the time, but have you ever thought about what types of illnesses spread through your schools and communities, and when they occur? The role of public health is to connect all those dots. Did you know that you, too, can connect those dots as a kid? You don't have to be an adult with a whole lot of degrees to do that. You just have to know the right questions and who to ask.

- **Ask and explore some health-related questions in your school.**
  - ☐ Survey your classmates about how they are feeling. Do they have specific symptoms of sickness? If so, when did they begin? By doing this, you can begin to figure out when flu season starts in your very own classroom.
    - Figure out how much influenza is going around your classroom, and when you might be more likely to get sick.
    - How many people have certain illnesses?
      - ☐ You can partner with your school nurse to use school as an opportunity to learn about outbreaks in your community.
- **Consider your own health compared to others.**
  - ☐ Jot down information about yourself and start to poll your experience and your classmates' experiences.
    - When you feel sick, how are you feeling (tummy ache, runny nose, etc.)?
      - ☐ What other ways might someone feel when they are not feeling well?
    - What did you eat before feeling sick?
- **Explore the health information in your community.**
  - ☐ Everybody lives in a county, and most counties have a health department. It's that health department's job to collect and report on patterns of disease in the community.
    - Steps to take:
      1. Figure out where your county's health department is and its contact information.
      2. Check out your county health department's website. What health and disease news has it reported on lately?
         a. Can you find information on how many cases of flu were reported in the last month? Foodborne disease? Measles?

3. Call the health department, say that you are a student interested in epidemiology, and ask:
   a. Are we seeing a lot of influenza? Where?
   b. What are some local patterns of illness happening within our community right now?
   c. What are some things the community can do to stop the spread of those diseases?

You might be surprised that your local health department is happy to talk with you!

resources and tools to organizations in need. She even worked at the Centers for Disease Control and Prevention (CDC) during the 2014 Ebola epidemic and the 2020 coronavirus (COVID) pandemic! In addition to helping the United States navigate the pandemic, another major challenge Dr. Beer had to overcome happened when she was very young. She was born blind due to cataracts in her eyes, and grew up having to wear contacts and glasses her whole life, which made her feel like an outsider.

Knowing what it felt like to be different helped Dr. Beer to become resilient and gave her a strong sense of justice that motivates much of what she has done in the past and present! Her work within the areas of epidemiology and public health has assisted in informing the field about diseases and impacted the health of individuals all across the globe!

The COVID-19 Pandemic

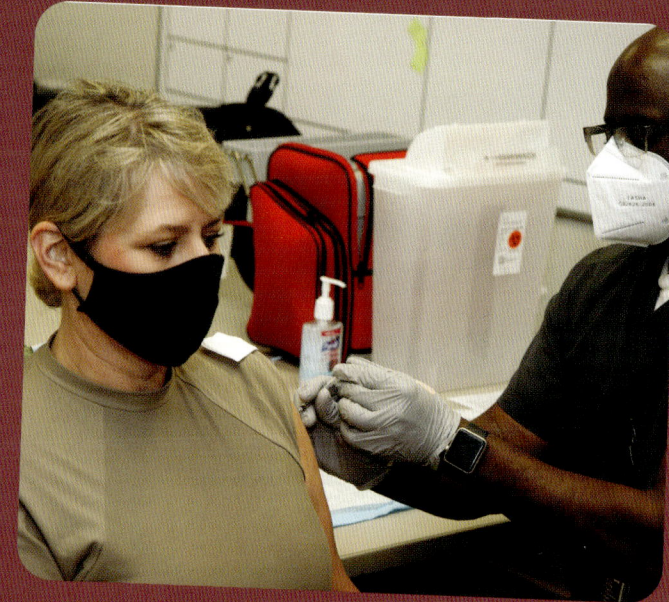

If you were alive in 2020, you most likely were affected in some way by the coronavirus known as COVID-19. Perhaps you or a loved one developed the disease, or maybe you had to wear a mask or shelter in place instead of going to school in person. The COVID-19 disease spread throughout the globe, causing a public health emergency of international concern, and ultimately resulting in more than three million deaths, according to the World Health Organization. The respiratory COVID-19 disease was caused by a virus called SARS-CoV-2, which led to a number of symptoms that ranged from mild to very severe. And like many viruses, SARS-CoV-2 could change and mutate, leading to new variations of the virus and illness it caused. However, thanks to scientists and medical professionals across the world who treated patients, developed vaccinations, and informed the population of what to do to prevent the illness, the world was able to get back to a new normal after the pandemic eased. This is just one example of the importance of science, research, public health systems, and the exceptional professionals who help keep the population healthy!

Science Activities

Medical Sciences

Activity 1:
Perception Puzzles

Activity 2:
Heart Rate

Activity 3:
Cataracts

Perception Puzzles

Neuroscientists help us to understand how our brains work, the relationship between our bodies and brains, and what we could do to improve our health and the function of these systems. To move through the world, we rely on proprioception, also known as kinesthesia, the body's ability to sense movement, action, and location. Neuroscientists perform tests to better understand how these abilities work. Some of these tests try to trigger a natural response to stimuli, something that excites your senses. You have likely experienced similar responses in your daily life that affect your perception or understanding of the world around you.

Using the activities below you will try to "trick" your brain and trigger some fun responses that may alter your perception of your environment.

Try to keep track and compare your responses as you go.

3-D Hands

1. Trace your hand on a white piece of paper with a marker or crayon. You could also laminate the paper or slide your hand into a reusable sheet protector and use dry-erase markers. If not reusing, any marker or crayon will work.

2. Using any colors you would like, create a pattern similar to what you see here.

3. The lines should be curved inside the outline of the hand, and straight on the rest of the paper. Once the paper is completely covered with stripes of color, the image should appear as 3-D even though it is only in 2-D.

X Marks the Spot?

1. Take a piece of construction paper and draw an X on it at least 2 inches in from each side. If you wish to repeat the experiment using the same page, you could laminate the paper or put your hand in a plastic sheet protector and use a dry-erase marker only.

2. Using your nondominant hand (the hand you don't write with), raise the paper above your head. Pick up a marker in your dominant hand, raise it above your head and attempt to make a dot on the paper as close to the X as possible. Look at your results. Try it again and see if you can get even closer. Switch hands and try it again—how did it go?

How would you compare the first, second, and third attempts?

Finding Your Fingertips

Raise both hands above your head. Keep the fingers of your left hand as still as you can. Close your eyes. Quickly touch the tip of your right index finger to your nose and then to the tip of your left thumb . . . then your nose, then your left index finger, your nose, then your left middle finger, your nose, then your left ring finger, your nose, and finally your left pinky finger.

- Remember to try to keep your left hand still!
- Quickly repeat the process with each finger on your right hand.
- What if you switch hands and try again?

What's Happening?

If we can't see or hear, we often rely on the sense of touch in our extremities (hand/fingers and feet/toes) to tell us where we are and what is going on around us.

For most people, it is hard to fight that urge to wiggle your fingers in the 3-D Hands experiment. You may notice that with repetition, X Marks the Spot and Finding Your Fingertips become easier.

Removing access to sight forces you to rely on your sense of touch in a new way. Every time you repeat an activity, you provide your brain with more data or information; this will affect your perception of the world. Our navigation of the world often improves with more information.

Activity 2

Heart Rate

Our cells all have jobs to do. To perform these jobs, cells use oxygen to release energy. When activity levels are higher than normal, more oxygen is required. Your heart and lungs work together to deliver all the oxygen your body needs. Exercise and an active lifestyle make you feel good, and they also help keep your heart healthy. This is important, because cardiovascular disease (heart disease) is the leading cause of death in the United States.

Calculate your resting heart rate, then check your heart rate after being active. Create a data table and bar graph to compare your body's response to different physical activities.

Materials Needed:

- Timer (stopwatch or app)
- Paper
- Pencil
- Exercise equipment (optional)
- Ruler (optional)
- Graph paper (optional)

The Experiment

1. Calculate your resting heart rate—that's how fast your heart beats when you're relaxed. Have a seat and take a few deep breaths to get comfortable before starting.

- Find your pulse in your wrist or neck.
- Count the number of beats you feel in 15 seconds. Use a timer to help with this.
- Multiply that number by 4. That is your resting heart rate in beats per minute.

2. Choose some exercises or activities you enjoy. Examples include walking, lifting weights, doing jumping jacks, playing basketball, and dancing. Choose at least three activities.

3. Create a data table to record changes to your heart rate during physical activity. Your data table should have columns for recording your heart rate and rows for different activities.

My Resting Heart Rate:	
ACTIVITY	HEART RATE AFTER 5 MINUTES
Activity 1:	

Activity 2:	
Activity 3:	

4. Collect heart rate data. Choose one of the activities to start with.
   Tip: Before starting, take a moment to stretch!

   - Set the timer for 5 minutes and start your activity.
   - When the timer goes off, stop the activity and calculate your heart rate like you did earlier. Record it on the data table.
   - Continue the activity if you're having fun!

5. Take a break and let your heart go back to its resting rate. Pick another activity and collect data just like before. Repeat for the third activity.

Once you have collected data from all three activities, look at the numbers. What do you notice? Did your heart rate go up or down? Now share your findings with your family or friends.

Activity 3
Cataracts

Your eyes are made of many parts that need to stay healthy for you to see well, including the cornea, iris, retina, optic nerve, and lens. Injury, diseases, and aging of these parts can cause changes to a person's vision. The lens focuses light coming into your eye to create an image. Cataracts are a disease that causes the lens to become cloudy, causing blurry vision and even blindness. Below are two images; the one on the left is what a person without cataracts would see. The image on the right is what you might see with cataracts.

Images courtesy of
Kamin Science Center.

Materials Needed:

- A pair of old sunglasses
- Clear plastic sandwich bags
- Markers
- Scissors
- Tape
- Paper

Create special glasses to simulate blurry vision and test your eyesight!

The Experiment

1. Pop the lenses out of an old pair of sunglasses (make sure you ask the owner's permission first!). Use the frames to make blurry vision glasses.

- Use a marker to trace the shape of the lens on the sandwich bags.
- Cut several new lenses out of the sandwich bags.
- Tape a new lens onto each side of the sunglasses' frame.
- Make sure you cut extra lenses from the plastic sandwich bags—you'll need them later!

2. Next, you'll need something to test your vision with the blurry glasses.

- Write a message (or random letters) on paper in large letters. Or you can cut a picture out of a magazine or look at a photo on the wall.
- Tape your message up on the other side of the room.

3. You will look through your blurry glasses at the message or picture you placed on the other side of the room. Then you'll add more lenses to the glasses and observe how your vision changes. Create a data table to record your observations.

Number of Lenses	Observations
1	
2	
3	
4	
5	
6	

4. Sit in a chair or stand about 10 feet away from your message.

5. Put on the glasses and look at your message. Write down your observations in the data table.

6. Add additional lenses to each side of your glasses, then look at your message again from the same spot. Make more observations on your data table.

7. Continue adding more lenses and making observations until you can no longer see your message.

- How many lenses does it take to make it difficult to read your message?
- How many lenses does it take to completely block your vision?

Explore More!

Opportunities to Explore TODAY

Opportunities to Explore LATER

Attending College

Degree Pathways

Scientific Pathways

What Else Can a Scientist Do?

Resources and References

Ways to Explore Science TODAY and LATER

Opportunities to Explore TODAY

Did you know that there are a number of avenues and programs aimed at getting kids just like you into science TODAY? Here are some opportunities you can explore at your school, in your community, or in programs nearby, that may get you started on your path to the beautiful world of science. Feel free to ask your teacher for help!

Science Fairs

Science fairs are competitions in which students showcase science-related investigations and projects they participated in and conducted. Students conduct their own research on a specific topic and utilize the steps of the scientific method or engineering designs to run their experiments. Then, they create posters and presentations displaying their findings, and share these at science fair events.

Examples: Science fairs can be held at school, city, state, national, and global levels.

Volunteer Opportunities

Volunteering or interning means providing services without pay. People can get important hands-on experiences at different places within their communities where science or science-related duties are conducted. Although these are unpaid opportunities, they often provide training and build knowledge.

Examples: Volunteer at an animal shelter if you like animals, or check out opportunities at planetariums or public gardens to explore space or nature.

Youth Science Programs

Youth programs are activities that students can participate in related to different areas of interest, including science! These programs are often offered at sites within the community or at schools and include meetings where participants engage in fun and educational activities.

Examples: 4-H for those interested in agriculture and other scientific areas, or science-related summer programs and after-school programs.

Science Competitions

Science competitions are events where youth from all over come together and compete in scientific challenges for prizes. Youth spend many months practicing and preparing for competition and then showcase their knowledge and work at these large events. Prizes can span from certificates and ribbons to medals and scholarship funds.

Examples: The National Science Bowl, First Tech Challenge, Zero Robotics, and International Chemistry Olympiad

Opportunities to Explore LATER

Here are some opportunities you can explore after middle school, but especially in high school and college. Counselors, teachers, career center staff, and community members as well as searches on the internet may help you find some of these opportunities near you!

Scholarships

A scholarship includes funds/money awarded to a student to help support their education. Usually, students must meet specific criteria to be eligible for scholarships, and often must complete an application and submit documents to show their academic histories and abilities, such as transcripts or personal statement papers. Scholarships can be offered from a variety of sources including schools, community programs, and even the government.

Example: Gates Millennium Scholarship Program

Fellowships

A fellowship is similar to a scholarship as it offers funds/money to students to help support their education, and students must meet criteria and often need to apply for the funds. However, fellowships are often offered to students who have already completed their undergraduate work, and the funds may be available for a longer period of time (several semesters, years, etc.).

Example: The National Science Foundation Graduate Research Fellowship Program

Assistantships

An assistantship is an employment position students can take that helps pay for their education. Assistantships require students to work while they complete their degrees and will help waive some of the cost of college while often providing additional funds for living expenses.

Examples: Research assistants can help college professors conduct research, while teaching assistants can help professors teach a class.

Attending College

One of the exciting things about exploring the fields of science is the opportunity to learn as much as you can—and attending college can help you do this! Many careers in science require degrees to help you develop expertise, to get experience within general and specific fields of science, and to prepare you for your selected science-related field.

College Majors

A college major is an area of study in which a student completes coursework and other requirements. Majors include a range of classes that give students important information, knowledge, and theory in their chosen subject. Majors generally include major-related coursework, general education requirements, and any other college or university requirements.

College Minors

A college minor is an optional area of study that can be completed alongside a college major. Minors give students a more specialized glimpse into another area of study, providing more information than non-minors receive, but less information than a student majoring in that subject. For example, a student can major in psychology, which covers humans from before birth through death, and minor in developmental psychology or child development, which focuses on children from before birth to around age eighteen.

Postdoctoral Positions

A postdoctoral position, also called a postdoc, is a position that a student can attain after graduating with their doctoral degree. Postdocs allow students to get more experience and training in specialized scientific areas and to continue working with college professors and researchers to conduct research. Postdocs are often paid positions and come with many benefits, such as helping students publish articles.

Example: A postdoc in developmental psychology

Degree Pathways

Associate Degree
2 Years

> **EXAMPLES:** Associate of Arts (AA) or Applied Arts (AAA), Associate of Science (AS) or Applied Science (AAS), and Transfer Degrees

Associate degrees are *undergraduate* degrees that allow students to engage in more coursework after graduating high school. These degrees take an average of two to three years and can be completed at community colleges. They often include the completion of general education classes as well as major classes.

Bachelor's Degree
4 Years

> **EXAMPLES:** Bachelor of Arts (BA), Bachelor of Science (BS)

Bachelor's degrees are *undergraduate* degrees that expose students to a wide range of subjects as well as specialized information related to their field of study. They include the completion of general education classes and major and minor classes. These degrees take an average of four years (can be fewer if students have an AA and/or other transfer credits) and can be completed at colleges and universities. Bachelor's degrees provide students with knowledge and practical skills within their selected fields, and often help graduates be more competitive in job markets than individuals who do not have these degrees.

Graduate and Professional Degrees
6–8 Years

> **EXAMPLES:** Master of Arts (MA) or Science (MS), Doctorate of Philosophy (PhD), Doctorate of Medicine (MD), etc.

After completing a bachelor's degree, students are considered graduates and can continue on to graduate school, where they focus more narrowly on their subject of choice. These programs tend to be highly specialized, meaning students get a much more in-depth understanding of their field, as well as more practice using tools and methods needed within their field. Generally, they take more coursework, but must also complete practical courses where they do hands-on tasks within their field, and must complete examinations, projects, papers, and/or studies to show their understanding of the materials learned. These degrees take an average of four to eight years, depending on the area of study and must be completed at universities.

After completing these degrees, students are expected to have expertise within their field. Some fields, such as medicine or teaching, require additional training, such as residencies and student teaching, after students complete their required coursework and examinations.

Scientific Pathways

There are many ways to enter the science fields and a host of resources that can help. Some are more common than others. Here are a few helpful opportunities and questions to consider along the pathways shared by the scientists in this book.

Internships and Volunteering

Some young scientists participate in internships to learn more about their field. This gives them hands-on experience within the field. If you know of an area of study you are interested in, see what types of volunteer opportunities and internships you can participate in. This will give you direct experience in your science of interest and can be a great addition to your list of extracurricular activities, which may help you be eligible for scholarships and can help you with your college applications!

Corina Newsome allowed her passion for animals to lead her to a veterinarian's office, where she volunteered for many years. That position helped her know that she loved working with animals, but not necessarily in a medical environment. Then, her internship at a zoo gave her even more experience and set her on her way to pursuing biology in college and to finding a career that she is really good at and enjoys as well!

Community College

Community colleges are college institutions that give students a chance to complete two-year degrees, such as associate degrees, as well as diplomas and certificates. Individuals from the community can register to become students, and then complete a number of courses and other major-related requirements to finish their programs. Once students complete their programs, they can move into the workforce or continue their educational studies and transfer to four-year colleges or universities.

After Dr. Vernard Lewis graduated high school, he went to a community college, which allowed him to build more skills and learn more about his field of interest. Then he moved on to UC Berkeley, where he was able to finish his bachelor's, master's, and doctorate degrees! Community colleges are great midway points for those interested in pursuing careers in science!

Work/Life Balance

Scientific inquiry can often take up a lot of time, as scientists are engaged in long hours of studying concepts, developing methods, conducting experiments, analyzing the results, and then sharing their research. A single experiment can span a few months or several years. Given these demands, many scientists strive to find a balance between their work and non-work lives. Some choose to get married and start families; most spend time with friends or engage in activities like athletics, games, or the arts. Achieving a balance between work and home can be an important and challenging goal. But to live full, fulfilled lives, finding love, pursuing non-job-related hobbies and interests, and caring for others can often help professional scientists attain meaning beyond their work.

Dr. Alison Gopnik got married and had several children as she navigated her way through science. Despite the challenges of balancing a family and an academic career, she was able to do this in a fulfilling way. Moreover, she advocates for better systems within academia to allow young professionals to not only be successful in their scientific jobs, but also to have the space in their lives to sustain healthy families.

What Else Can a Scientist Do?

Although we often think of scientists as being busy in laboratories or college classrooms, scientists can also engage in a wide range of other endeavors during their careers. Here are some additional things a scientist can do!

INVENT

Scientists can take their knowledge and curiosity and create something brand-new that has never been created before. This is called an *invention*. Inventions can be items, devices, methods, processes, etc that do something new, or they can build on previous inventions. Once someone has created a new invention, they can apply for a patent to protect their ownership over their invention.

Dr. Patricia Bath created her Laserphaco Probe, which was a revolutionary device used to help treat cataracts. Although it was groundbreaking at that time, Dr. Bath knew that new inventions and technologies were being created all the time, and that was important for the advancement of science. Today, newer technologies and methods are used to treat cataracts, but thanks to Dr. Bath's early contributions, newer technologies built on her work and the work of others who came before and after her.

SERVE ON AN ADVISORY BOARD

An advisory board is a group of individuals who come together to discuss and solve problems that can impact organizations and communities, and those affected by them. Scientists can serve on advisory boards to offer important knowledge related to their field, which can help others make insightful and informed decisions.

> Dr. Pedro Sanchez served on advisory boards to a US president and other world leaders to provide his expertise to help address difficult problems related to climate, environment, and agriculture.

DEVELOP SYSTEMS AND SCALES

Within all different types of disciplines and industries, determining the best ways to conduct methods and processes, and then evaluating the outcomes, is important. For example, understanding which skills are needed to teach a child how to read, using a process or method to teach them, and then testing whether the child developed those skills could be considered a system. Scientists can use their knowledge about their areas to create effective systems and also to determine whether those systems are doing what they are supposed to do by developing measures and scales to evaluate the systems and the outcomes they produce.

> Dr. Temple Grandin used her knowledge about animal behavior and her ability to think outside the box to develop facilities that serve animals in a humane way, but also to use systems to audit whether the methods and processes were effective. She was able to create systems and scales related to animal welfare, and then work to make sure these were implemented in large companies to better serve the animals they maintain, and the community they impact.

WRITE AND PUBLISH

Scientists often have the chance to write about science and their work. They can publish articles, books, reports, and so much more that allow them to tell others about what they have learned, what they know, and how to use that information in meaningful ways.

> W. E. B. Du Bois spent a large part of his academic career writing books, articles, essays, and literature supporting the Black community. He even created *The Crisis*, a publication that was a messenger for the NAACP. Additionally, he published many books including *The Souls of Black Folk*, which catapulted his activism and made him a spokesperson for the Black community.

EDUCATE AND ADVOCATE

Scientists can often tour the world and share their knowledge within a variety of venues and media outlets. They can advocate for change and educate the public on matters that are important to them.

> A part of what Mariah Gladstone does is speak to the public and share her knowledge, recipes, and cooking tips through her videos as well as her appearances on television programs, such as the *Today* show. In addition, she has used her platform to share her knowledge at large conferences like those held by the Smithsonian's National Museum of the American Indian and organizations such as the Montana Cooperative.

TRAVEL THE WORLD

Although scientists spend a fair amount of time doing their research, conducting their experiments and studies, and writing up their findings, they also get the chance to share their work with others. They can travel all over the world to present their work, to consult with others, to collaborate, and to use their work to make a better world.

> Dr. Michael Blakey travels the world to talk with other scientists about ethical practices related to human bones. He wants to make sure that the remains of all humans are treated with respect and dignity, and that the communities who were family and ancestors of those individuals feel satisfied with how the bones and remains are handled—no matter where they are found!

WRITE BOOKS FOR CHILDREN

Scientists can write books for children! They can use their scientific knowledge and include it in engaging books for young readers to read and learn new things.

> Dr. Rajani LaRocca has published several books, many exploring various scientific topics such as DNA (*The Secret Code Inside You*, Little Bee Books), the human heart (*Your One and Only Heart*, Dial Books), vaccinations (*A Vaccine Is Like a Memory*, Simon & Schuster), and her Newbery Honor winner about human blood and leukemia (*Red, White, and Whole*, Quill Tree Books).

WORK IN NONPROFIT ORGANIZATIONS

A nonprofit organization is a business that serves a purpose other than making money. Nonprofit organizations work to make change in their communities, and any money raised goes back into the organization to further its cause. Scientists can form nonprofit organizations, or help preexisting nonprofits, if they want to play a bigger part in serving communities or addressing problems they see and care about.

> The Jane Goodall Institute is a nonprofit organization that strives to educate the public about the importance of wildlife conservation as well as work to conserve the natural world and those who inhabit it. You can learn more about this organization at JaneGoodall.org.

Well, young scientist, you have successfully made it to the end of this book! Way to go!

Hopefully the information included within—the stories shared about challenges and triumphs of amazing scientists, the questions asked and experiments posed, and the vast opportunities to explore science and the pathways to scientific careers—has left you excited about your own scientific story. You are on your way!

In closing, I ask you, **what** questions will YOU ask and **why** are they important to YOU? **Which** disciplines will YOU explore further and which methods will YOU employ? **How** will YOU gather data and **who** will YOU share your findings with? **Where** will YOU conduct your own research and in what ways will YOU help the world with your impressive and groundbreaking scientific discoveries? **When** will YOU begin? (I hope you answer "RIGHT NOW!")

With whatever pathways you decide, remember that there will be challenges—but do not let them stop you. Use the remarkable stories of these twenty-five trailblazing scientists as inspiration to help you blaze your own amazing trail.

Now go out there and explore, young scientists!

The world is waiting for you!

References and Resources

Adero, Malaika. 2020. *A Black Woman Did That*. New York: Downtown Bookworks.

Albee, Sarah, and Gustavo Mazali. 2020. *Jane Goodall: A Champion of Chimpanzees. I Can Read!* New York: HarperCollins.

Allen, Nicola J., and Richard Daneman. 2018. "In Memoriam: Ben Marres." *Journal of Cell Biology,* 217 (2): 435–438. https://doi.org/10.1083/jcb.201801019.

American Medical Association. 2023. "PolicyFinder: Definition of Surgery H-475.983." https://policysearch.ama-assn.org /policyfinder/detail/surgery?uri=%2FAMADoc%2FHOD .xml-0-4317.xml.

Barres, Ben. 2018. *The Autobiography of a Transgender Scientist*. Cambridge, MA: MIT Press.

Bedi, Joyce. 2021. "Diverse Voices: Women Inventors." Lemelson Center for the Study of Invention and Innovation. March 22, 2021. https://invention.si.edu/invention-stories /diverse-voices-women-inventors

Brigida Family. "An American Portrait: Kurzweil 250 CBS News - 1984," posted December 12, 2015, YouTube video, 15:41. https://www.youtube.com/watch?v=lyAI5b8jQfY.

Britannica, The Editors of Encyclopedia. "*Cryptography*." *Encyclopedia Britannica*, December 6, 2023. https://www .britannica.com/topic/cryptography.

Britannica, The Editors of Encyclopaedia. "*Surgery*." *Encyclopedia Britannica*, February 1, 2024. https://www .britannica.com/science/surgery-medicine.

Brown-Wood, JaNay. May 30, 2023. Personal Interview with Omar Lizardo.

Brown-Wood, J. May 31, 2023. Personal Interview with Temple Grandin.

Brown-Wood, J. June 1, 2023. Personal Interview with Pedro Sanchez.

Brown-Wood, J. June 9, 2023. Personal Interview with Vernard Lewis.

Brown-Wood, J. June 19, 2023. Personal Interview with Ellen Ochoa.

Brown-Wood, J. June 20, 2023. Personal Interview with Corina Newsome.

Brown-Wood, J. June 28, 2023. Personal Interview with Kay Savage.

Brown-Wood, J. June 28, 2023. Personal Interview with Mariah Gladstone.

Brown-Wood, J. June 29, 2023. Personal Interview with Alison Gopnik.

Brown-Wood, J. June 29, 2023. Personal Interview with Rajani LaRocca.

Brown-Wood, J. July 18, 2023. Personal Interview with Kay Savage.

Brown-Wood, J. July 18, 2023. Personal Interview with Terence Tao.

Brown-Wood, J. July 19, 2023. Personal Interview with Ray Kurzweil.

Brown-Wood, J. July 21, 2023. Personal Interview with Natasha Tilston-Lunel.

Brown-Wood, J. October 2, 2023. Personal Interview with Nadine Caron.

Brown-Wood, J. November 10, 2023. Personal Interview with Karlyn Beer.

Cairns, Chris. "Kurzweil - It All Started with Ray (The Kurzweil Music Story)," posted January 14, 2022, YouTube video, 10:41. https://www.youtube.com/watch?v=HpfPyOD1Gks.

Calvert, Jennifer, and Octavia Jackson. 2021. *Science Superstars: 30 Brilliant Women Who Changed the World*. New York: Castle Point Books.

Centers for Disease Control and Prevention. December 9, 2022. "ASD Diagnosis, Treatment, and Services." https: //www.cdc.gov/autism/about/?CDC_AAref_Val=https: //www.cdc.gov/ncbddd/autism/facts.html

Centers for Disease Control and Prevention. September 1, 2023. "SARS-CoV-2 Variant Classifications and Definitions." https://www.cdc.gov/coronavirus/2019-ncov /variants/variant-classifications.html.

Corrigan, Jim. 2009. *Profiles in Mathematics: Alan Turing*. Greensboro, NC: Morgan Reynolds Publishing.

Cunningham, Meghan Engsberg. 2016. *W. E. B. Du Bois: Co-Founder of the NAACP*. Great American Thinkers. New York: Cavendish Square Publishing.

Dalal, Anita. 2022. *Improving Health: Women Led the Way.* Super Sheroes of Science. New York: Scholastic.

Du Bois. W. E. B. *Darkwater: Voices from Within the Veil* (New York, 1920; Project Gutenberg, 2005). https://www.gutenberg.org/files/15210/15210-h/15210-h.htm.

Edwards, Roberta. 2012. *Who Is Jane Goodall?* New York: Penguin Random House.

Fallik, Dawn. 2022. "Scientist Finds Professor Who Supported Her Love for Bugs When She Was 4." *Washington Post*, May 21, 2022. https://www.washingtonpost.com/science/2022/05/21/bugs-rebecca-varney-vernard-lewis/.

Gallagher, Ashley. 2018. "Eugenie Clark – The Shark Lady." Smithsonian National Museum of Natural History." https://ocean.si.edu/ocean-life/sharks-rays/eugenie-clark-shark-lady.

Genzlinger, Neil. 2017. "Ben Barres, Neuroscientist and Equal-Opportunity Advocate, Dies at 63." *New York Times*, December 29, 2017. https://www.nytimes.com/2017/12/29/obituaries/ben-barres-dead-neuroscientist-and-equal-opportunity-advocate.html.

Gregersen, Erik. "*Prime Numbers*." *Encyclopedia Britannica*, November 24, 2016. https://www.britannica.com/story/prime-numbers.

Hardy, P. Stephen, and Sheila Jackson Hardy. 2000. *Extraordinary People of the Harlem Renaissance.* New York: Children's Press.

Henderson, Harry. 2011. *Alan Turing: Computing Genius and Wartime Code Breaker.* Makers of Modern Science. New York: Chelsea House Publishers.

Huberman, Andrew D. 2018. "Ben Barres (1954–2017)." *Nature* 553 (7688): 282. https://doi.org/10.1038/d41586-017-08964-1. PMID: 32094595.

Jackson, Tom. 2011. *DK Eyewitness Books: Science.* London: DK Publishing.

Johnson, Katherine. 2019. *Reaching for the Moon: The Autobiography of NASA Mathematician Katherine Johnson.* New York: Atheneum Books for Young Readers.

Johnson, Katherine, Joylette Hylick, and Katherine Moore. 2021. *My Remarkable Journey: A Memoir.* New York: Amistad.

Keating, Jess, and Marta Álvarez Miguéns. 2017. *Shark Lady: The True Story of How Eugenie Clark Became the Ocean's Most Fearless Scientist.* Naperville, IL: Sourcebooks Jabberwocky.

Kids Britannica, The Editors of Encyclopedia. n.d. "*Eugenie Clark*." *Kids Encyclopedia Britannica*. https://kids.britannica.com/kids/article/Eugenie-Clark/633305#.

Lang, Heather, and Jordi Solano. 2016. *Swimming with Sharks: The Daring Discoveries of Eugenie Clark.* Chicago: Albert Whitman & Company.

Lawlor, Laurie. 2017. *Super Women: Six Scientists Who Changed the World.* New York: Holiday House.

Lord, Michelle, and Alleanna Harris. 2020. *Patricia's Vision: The Doctor Who Saved Sight.* New York: Sterling Children's Books.

McFadden, Robert D. 2015. "Eugenie Clark, Scholar of the Life Aquatics, Dies at 92." *New York Times*, February 25, 2015. https://www.nytimes.com/2015/02/26/us/eugenie-clark-scholar-of-the-life-aquatic-dies-at-92.html.

Meltzer, Brad, and Chris Eliopoulous. 2016. *I Am Jane Goodall.* Ordinary People Change the World. New York: Rocky Pond Books.

Moitra, Ilina. 2021. "Ben Barres: An Ode to the Man Himself." *Kinesis Magazine*, April 12, 2021. https://kinesismagazine.com/2021/04/12/ben-barres-an-ode-to-the-man-himself/.

Mosca, Julia Finley. 2017. *The Doctor with an Eye for Eyes: The Story of Dr. Patricia Bath.* Amazing Scientists 2. Seattle: The Innovation Press.

Mote Marine Laboratory & Aquarium. n.d. "Meet the Team: Dr. Eugenie Clark." https://mote.org/staff/member/eugenie-clark.

Nagelhout, Ryan. 2017. *Alan Turing: Master of Cracking Codes.* New York: Rosen Publishing Group.

National Aeronautics and Space Administration. 2017. "Katherine G. Johnson." May 25, 2017. https://www.nasa.gov/people-of-nasa/katherine-g-johnson/.

National Environmental Satellite, Data, and Information Service. n.d. "Learn about Soil Types." https://www.nesdis.noaa.gov/learn-about-soil-types.

National Marine Sanctuary Foundation. 2021. "Celebrating Wave Makers for Women's History Month: Dr. Eugenie Clark 'The Shark Lady.'" https://marinesanctuary.org/blog/celebrating-wave-makers-dr-eugenie-clark/.

National Ocean Service/National Oceanic and Atmospheric Administration. n.d. "Dr. Eugenie Clark (1922–2015): The Life and Legacy of an Ocean Pioneer." https://oceanservice.noaa.gov/news/may15/eugenie-clark.html.

Robeson, Teresa. 2019. *Queen of Physics: How Wu Chien Shiung Helped Unlock the Secrets of the Atom*. New York: Sterling Children's Books.

Rudwick, Elliott. *W. E. B. Du Bois*. Encyclopedia Britannica, December 28, 2023. https://www.britannica.com/biography/W-E-B-Du-Bois.

Sánchez Vegara, Maria Isabel, and Beatrice Cerocchi. 2018. *Little People, Big Dreams: Jane Goodall*. London: Frances Lincoln Children's Books.

Sánchez Vegara, Maria Isabel, and Ashling Lindsay. 2020. *Little People, Big Dreams: Alan Turing*. London: Frances Lincoln Children's Books.

Schwartz, Ella. 2021. *Stolen Science: Thirteen Untold Stories of Scientists and Inventors Almost Written Out of History*. New York: Bloomsbury Children's Books.

Schwartz, Heather E. 2018. *Code-Breaker and Mathematician Alan Turing*. STEM Trailblazer Bios. Minneapolis: Lerner Publications.

Silvey, Anita. 2015. *Untamed: The Wild Life of Jane Goodall*. Washington, DC: National Geographic Kids.

Simon, Seymour. 2012. *Science Dictionary*. Mineola, NY: Dover Publications.

Slade, Suzanne. 2019. *A Computer Called Katherine: How Katherine Johnson Helped Put America on the Moon*. New York: Little, Brown Books for Young Readers.

Tao, Terence. 2020. *Vitae and Bibliography for Terence Tao*. Last updated October 16, 2020. https://www.math.ucla.edu/~tao/preprints/cv.html#.

Techmoan. "Kurzweil's Revolutionary Reading Devices for the Blind," posted May 29, 2021, YouTube, 12:40. https://www.youtube.com/watch?v=g0jECuwrn_U.

Tempoli, Marzia, ed. 2020. *Science*. New York: Smartbook Media.

Thimmesh, Catherine. 2022. *Girls Think of Everything: Stories of Ingenious Inventions by Women*. New York: Clarion Books.

Tilston Lab. n.d. "What Are Bunyaviruses?" https://www.thetilstonlab.org/about-5.

Time. n.d. *The Inventor: Patricia Bath*. https://time.com/collection/firsts/4898565/patricia-bath-firsts/.

Tran, Kathalina. 2017. "Voyager 1 Trajectory Through the Solar System." Updated November 15, 2023. https://svs.gsfc.nasa.gov/4139.

Troy, Don. 2010. *W. E. B. Du Bois. Journey to Freedom*. Mankato, MN: Child's World.

U.S. National Library of Medicine. 2015. "Changing the Face of Medicine: Dr. Patricia E. Bath." Updated June 3, 2015. https://cfmedicine.nlm.nih.gov/physicians/biography_26.html.

Venezia, Mike. 2010. *Jane Goodall: Researcher Who Champions Chimps*. Getting to Know the World's Greatest Inventors & Scientists. New York: Children's Press.

We Are Tech Women. 2020. "Inspirational Quotes: Katherine Johnson, NASA Mathematician." October 3, 2020. https://wearetechwomen.com/inspirational-quotes-katherine-johnson-nasa-mathematician/.

Wheeler, Jill C. 2012. *Chien-Shiung Wu: Phenomenal Physicist*. Women in Science. Minneapolis: Checkerboard Library.

Wittrock, Jeni. 2015. *W. E. B. Du Bois*. North Mankato, MN: Capstone.

World Health Organization. n.d. "Coronavirus Disease (COVID-19)." https://www.who.int/health-topics/coronavirus#tab=tab_1.

World Health Organization. n.d. "The True Death Toll of COVID-19: Estimating Global Excess Mortality." https://www.who.int/data/stories/the-true-death-toll-of-covid-19-estimating-global-excess-mortality.